RISK ASSESSMENT
AND
DECISION MAKING
IN BUSINESS
AND INDUSTRY

A PRACTICAL GUIDE

GLENN KOLLER

CRC Press
Boca Raton   London   New York   Washington, D.C.

Koller, Glenn R. (Glenn Robert), 1951–
    Risk Assessment and Decision Making in Business and
Industry: A Practical Guide / by Glenn R. Koller.
        p.   cm.
    Includes bibliographical references and index.
    ISBN 0-8493-0268-4 (alk. paper)
    1. Risk assessment.   2. Decision making.   I. Title.
HD61.K63   1999
658.15′5--dc21   98-43655
                                                                                          CIP

No claim to original U.S. Government works
International Standard Book Number 0-8493-0268-4
Library of Congress Card Number 98-43655
Printed in the United States of America 1 2 3 4 5 6 7 8 9 0
Printed on acid-free paper

To Karen —
who has with love, good humor, and
understanding, supported me in this effort.
To Stephanie, Kathy, and Matthew —
It's over, I'm back!

Acknowledgments

Mike Long, Phil Hammond, and Cheryl Murry are three of the most intelligent, hard-working, and talented people I have had the pleasure to know and to work with. Mike's and Phil's dedication to our risk-software-development effort and educational assault are mainstays of our continued success. Cheryl's addition to the group has added mathematical prowess and sorely needed business experience. It truly is an honor and a privilege even to know, much less work with, such stellar individuals.

I acknowledge and thank Walt Reed, Terry Wood, and Bill Wishlinski for their continued support of our risk effort. Their tolerance for mavericks is appreciated, and Walt's patience and understanding are qualities every manager should possess.

I thank CRC Press editor Bob Stern for taking a chance on a guy who came from out of nowhere. Although publishing a book is a stressful situation, Bob's insightful guidance and ability to relate to people as a "regular guy" are qualities I have greatly appreciated. All of the staff at CRC Press have been great.

Last, but certainly not least, I express my deepest appreciation to my wife Karen, my daughters Stephanie and Kathy, and my son Matthew. Researching, writing, and publishing a book is an exceedingly time-consuming effort. Given my travel schedule, I had little enough time to give. During the production of this work, that little time was nearly completely usurped by the book-writing process. The family's technical support, encouragement, and good humor during this trying period are appreciated more than words can say.

About the Author

Glenn Koller is supervisor of Amoco Corporation's Risk Center of Excellence. He received his Ph.D. in geochemistry/geophysics from Syracuse University. Responsibilities of his current position include implementation of risk assessment and risk management technologies and processes in Amoco Corporation; development of statistical routines that comprise Amoco's risk systems; keeping abreast of risk technology developed by vendors, other companies, and the National Laboratories; marketing risk technology; and performing technical and consulting services. Areas of responsibility include business and product development, environmental concerns, ranking and prioritization of projects and products, and other aspects of Amoco's diversified business. Glenn currently resides in Tulsa, Oklahoma.

Contents

Chapter 1
Risk Assessment Advantages 1
Why Risk Assessment? 1
Consider Range of Possibilities 2
Establish Probabilities 3
Communication with Customers and the Public 3
Probabilistic Branching 4
Relationships Between Variables..5
Linking of Disciplines 6
Checklist — Don't Take It for Granted 7
Updating Estimates..8
To What Types of Problems Can Risk Assessment Be Applied? 9

Chapter 2
Risk Assessment — A Process..11
Risk Assessment — The Process 11
The Time It Takes to Do It Right and When to Get Started.....................14
Time for Preparation of the Organization — Different People See It
Differently 14
What to Provide...16

Chapter 3
Consistency — The Key to Risk Assessment.......................................21
Biggest "Bang" From Assessing Multiple Opportunities.........................21
"Type" Consensus Models 22
Common Measure...25
Risk Assessment Vision..26
Sometimes Better to Be Consistent Than to Be Right.............................28
Locking the Model 31
Users Will "Game" the System — The "Risk Police" 32
Databasing Inputs and Outputs 33

Chapter 4
Communication...35
Communication — Just One Shot...35
Communication — Why Bother?..36
Know Your Clients..36

Focus on Impact and Mitigation 37
Diplomatically Handle "Winners" and "Losers" 38
Communicating the Value of Risk Assessment 38

Chapter 5
Building a Consensus Model 41
What Is the Question? — Most of the Time and Effort 41
Consensus Model..42
Group Dynamics...44
Write It Down...45
Sort It Out..46
Group Dynamics Again 47
Units..47
Overarching Categories 49

Chapter 6
Build a Contributing-Factor Diagram 51
The Contributing-Factor Diagram — Getting Started....................................51
Identify and Define Variables 56
Ask the Right Question 57
Double-Dipping 58
Double-Dipping and Counting the Chickens..59
Fixing the Double-Dipping and Counting Chickens Problem 59
CFD-Building Example..60
Short List of Hints for Building a CFD...65

Chapter 7
Education — Critical at Three Levels 67
Education for Management 67
Introductory Class for Risk-Model Builders...69
Advanced Training for Risk-Model Builders...70
Organizational Attitude and Training Centers 71
Time Demands and Organization of the Education Effort.............................72

Chapter 8
Risk Assessment Technologies 77
Different Approaches...77
Discriminant-Function Analysis 77
Bayesian Analysis...79
Decision Trees 82
Factor Analysis 87
Neural Nets..89

Chapter 9
Monte Carlo Analysis 91
A Bit of History 91
For What Is It Good? 92
Simple Monte Carlo Example 92
How Many Random Comparisons Are Enough? 94
Output From Monte Carlo Analysis — The Frequency and Cumulative
Frequency Plots 95
Interpreting Cumulative Frequency Plots 100
Combining Monte Carlo Output Curves 104

Chapter 10
Decisions and Distributions 105
Decisions 105
Just What Is a Distribution? 106
Distributions — How to Approach Them 110
Symmetrical Distributions 112
Skewed Distributions 114
Spike Distributions 115
Flat Distributions 116
Truncated Distributions 117
Discrete Distributions 119
Bimodal Distributions 120
Reading Data From a File 121
Peakedness 123
Triangular Distributions 125
"Specific" Distribution Types 126

Chapter 11
Chance of Failure 129
Chance of Failure — What Is It? 129
Failure of a Risk Component 130
Chance of Failure That Does Not Affect an Input Distribution 133
Incorporating Chance of Failure in a Plot of Cumulative Frequency 134
Another Reason for Chance of Failure 136
The "Inserting 0s Work-Around" 138
Chance of Failure and Multiple Output Variables 139

Chapter 12
Time-Series Analysis and Dependence 141
Introduction to Time-Series Analysis and Dependence 141
Time-Series Analysis — Why? 141
Time-Series Analysis — How? 142

Time-Series Analysis — Results...143
Time-Series Analysis That's Not...144
Some Things to Consider...144
Dependence — What Is It?..147
Independent and Dependent Variables...148
Degree of Dependence..148
Multiple Dependencies and Circular Dependence....................................152
Effect of Dependence on Monte Carlo Output...153
Dependence — It's Ubiquitous...153

Chapter 13
Risk-Weighted Values and Sensitivity Analysis

Risk-Weighted Values and Sensitivity Analysis.......................................159
Introduction to Risk-Weighted Values and Sensitivity Analysis..............159
Risk-Weighted Values — Why?...159
Risk-Weighted Values — How?...161
The Net Risk-Weighted Value...164
The Economic Risk-Weighted Resource (ERWR) Value............................165
Risk-Weighted Values — The Answer...166
Sensitivity Analysis — Why?...168
Sensitivity Analysis — How?...170
Sensitivity Analysis — Results..172

Chapter 14
Risk Assessment Examples

Risk Assessment Examples...175
Introduction..175
Qualitative Model...175
Semi-Quantitative Model..177
Mini Time-Series Model...180
Fate-Transport (Process/Health) Model...187
Decision-Tree Conversion..192
Legal Model...198
Comprehensive Risk Assessment Example...204

Selected Readings..231

Subject Index...235

Introduction

PRACTICAL RISK ASSESSMENT

In the title of this book is the word "practical." That term will serve as the guiding principle for this entire volume. It seems the bulk of books with statistical or risk assessment themes are written for other people in the field, or to impress those same people. This book is intended to impress the reader with the facts that

- You don't need to be a statistical guru or risk assessment expert to use risk assessment in your decision-making process.
- Risk assessment can be practically applied at any decision-making level and in any business, educational, or technical arena.
- Risk assessment is a process — not just a set of equations — the success of which mainly hinges on skillful handling of organizational, educational, communication, and political aspects of the process. Technical considerations, while important, are secondary.
- Risk assessment can be explained and implemented using plain, understandable terms and concepts. Nebulous and (almost always) self-serving technical jargon is not necessary and is, in fact, detrimental.

In more than 20 years of working in this field, I have come to find that success, however defined, stems mainly from bringing to bear finely honed interpersonal, communication, and educational skills. Technical competence, while required, is an empty promise when not combined with the knack for engaging and facilitating individuals and groups.

It also has been my experience that many risk assessment processes touted by consultants tend to be a day's worth of process packed into a week. I find that people at all levels, in whatever business, are busy — busier perhaps than any person in past times with an equivalent position. Taking more time than is absolutely necessary to design a risk process and to implement the resulting risk model is a destruction of value. Hopefully, the practical approach put forth in this book will, at least in part, preclude such practices.

Like marriages in the U.S., a great proportion of risk assessment projects undertaken either fail outright or are unable to reach their intended goal or full potential. Risk assessment efforts tend to fail for many of the same reasons as failed marriages. Bad assumptions, poor communication, lack of sufficient partner/stakeholder intimacy, selfishness and arrogance, and unrealistic expectations are just a few of the foibles that can individually or in combination torpedo the most astute and technically brilliant risk assessment efforts. In this book I will address the technical aspects of risk assessments of all types as well as consider in depth the more difficult to grapple with organizational, political, and human issues.

Risk assessment is but part of the overall menagerie of processes that comprise the art and science of decision analysis. It is beyond the scope of this book to address any but the risk assessment-related aspects of decision making. Explanation of decision-making tools such as strategy tables and the like can be found in articles and books listed in the reference section for this introduction.

It is not the intent of this book to make risk assessment experts of those who now are not. Rather, the aim is to convey to the reader in plain English the fundamental principles of risk assessment and the enveloping processes. Although risk assessment software and "experts" still may need to be engaged to successfully implement a risk process in an organization, students of this book will understand risk-related processes and technologies. They will be able to intelligently discuss with those experts the unique considerations of their predicament and will be capable of interacting with the experts in the effort to solve the problem.

This book intentionally lacks a discipline slant. That is, in these pages you will find examples from, at least, the areas of

- Brokerage-house portfolio management;
- Legal decision making;
- Construction;
- Oil/gas exploration;
- Environmental assessments;
- Engineering;
- Marketing;
- Government;
- Manufacturing;

and many other areas. Risk assessment is a universally applicable technology and, as the following section will attest, it does not have to be hard to use.

YOU CAN DO IT!

As mentioned previously, it is my job every day to bring the risk process and risk models to bear on all types of businesses, processes, and situations. If I took the traditional approach to this problem — that of requiring of users things such as esoteric statistical information that they are not prepared to supply — I would fail miserably.

The key to successful risk-process and risk-model implementation is to build a risk system that requires from the user nearly nothing more than they presently can provide. If you expect that through zeal, fervor, osmosis, or a concentrated training effort you will convert all potential customers into rudimentary statistical beings, you are about to experience one of life's great disappointments.

When I think of a risk assessment, I do not have in mind only, for example, the engineer who is comfortable with spreadsheets, distributions, equations, and computer technology. Such individuals rarely can see how advanced they are in their thinking, abilities, and approach. They generally cannot appreciate why everyone else doesn't do it like they do it (what's the matter with these other people — this is easy!). To me, implementing a risk process means designing a system that requires from the user very little, if anything, that they presently are not prepared to deliver. This can be a challenge, but it is a challenge that I have successfully met time and again. You can too.

Most of this book is dedicated to explaining just how you might go about achieving this goal. Spending time with the potential users of the system, understanding their business, facilitating the risk-model-building process and many other steps and processes are required to build a successful risk model. The point is, however, that it can be done.

The result of a properly orchestrated risk-model-building process is a value-added risk model that is practical and comprehensive. As emphasized by the title of this book, "practical" is the operative term. If the system resulting from the risk-model-building effort is not practical and easy for the non-risk-expert to use, then the effort is a failure. Execution of the steps outlined in this book should lead to an easy-to-use risk model that actually can be a pleasure to work with. Using a risk model does not have to be hard!

1 Risk Assessment Advantages

CONTENTS

Why Risk Assessment? 1
Consider Range of Possibilities 2
Establish Probabilities 3
Communication with Customers and the Public 3
Probabilistic Branching 4
Relationships Between Variables 5
Linking of Disciplines 6
Checklist — Don't Take It for Granted 7
Updating Estimates 8
To What Types of Problems Can Risk Assessment Be Applied? 9

WHY RISK ASSESSMENT?

People are pressed into enacting risk assessments for various reasons. Foremost is the need to be able to consider a range of possibilities rather than a single-valued answer. Loosely speaking, a single-valued result is termed "deterministic" as opposed to a multi-valued (distribution-based) result that is referred to as "probabilistic." The term "probabilistic" is used because in addition to relating a range of possible values associated with the answer, a probabilistic result imparts the probability that values in given ranges will occur.

Another primary reason to perform a risk assessment is the uncertainty associated with many projects. For example, it may be that one of several properties is going to be purchased and the purchase price for each is fixed (no distribution of prices associated with each possible purchase). However, because of uncertainty associated with getting loans from lenders, the uncertainty that the properties will be ready for sale when you are ready to buy, the uncertainty related to the present owners being able to find other places to live, etc., you may want to deduce which investment is the most likely to succeed or fail. In this case, although you would be working with deterministic answers (fixed purchase prices), the probability or range of probabilities associated with each purchase may need to be known.

Risk assessment is also a very powerful "front end" for decision making and portfolio management. Although it is true that risk assessments sometimes are run on individual "stand-alone" situations, more commonly the aim of a risk assessment process is to assess several opportunities and to be able to compare, rank, and portfolio-manage the options. To do this, a robust risk model or models must be generated (models that can handle the varied aspects of all opportunities to be assessed), and a common measure must be agreed upon by which all opportunities can be compared.

For example, a company may have x dollars to invest. There are several investment opportunities available. Two are purchase opportunities: one a restaurant chain and the other several clothing stores. Two other opportunities are investment types. One investment is to provide start-up money for a new firm. The other is to invest in a particular international mutual fund. In this case, you likely would begin down the risk assessment road by deciding on what parameter you are going to use to compare these opportunities. Because they all involve the investment of dollars by the company, perhaps you would settle on determining the net present value (NPV) of each opportunity for comparison.

Having settled on NPV as the desired risk assessment result, it is likely that you would proceed to generate two models. One model might be designed to consider the aspects of purchasing another company. The model would need to be robust enough to be able to assess very different types of businesses — in this case, a clothing business and a restaurant business. This model, however, would generate as a result the NPV of each purchase. The second model would have to be comprehensive enough to handle different investment types. In this case, you are trying to compare a "seed money" commitment with a mutual fund investment. Again, like the business-purchase model, the desired output from the investment risk model is NPV.

Ultimately, these two models would result in four NPV distributions for comparison. The art (and sometimes science) of portfolio management is then applied to determine which opportunity or combination of opportunities will be pursued. In this way, risk assessment serves as an essential "front end" for decision making and portfolio management.

CONSIDER RANGE OF POSSIBILITIES

Lo, those many years ago when I was a graduate student, as part of a geochemical experiment I would solve a system of equations that calculated a value. I hoped the calculated value would predict the outcome of the actual physical analysis. The single calculated value or series of values resulting from re-solving the equations with slightly different coefficients did not afford me any "feel" for whether my results would be even remotely coincident with real-world values. That is, the values resulting from re-solution of the equations with various coefficients did yield a range of answers, but I could never be sure how much of the actual (real-world) range this represented. In addition, it was just too time consuming to even attempt to compute all of the possible permutations.

A well-constructed risk model alleviates this conundrum by using ranges of values to represent most, if not all, variables that constitute a set of equations. Application of Monte Carlo techniques (see Chapter 9) to the variables and equations will yield a range of answers. The Monte Carlo technique will cause all reasonable combinations of variable coefficients to be combined in equations. This will yield an output curve that represents all probable solutions.

Given the sophistication of modern risk assessment software, there is no reason to resort to discretization of distributions (attempting to represent a distribution with a small number of discrete values). Reduction of a distribution to "representative" discrete values drastically reduces the information content of a variable and thus is to be avoided.

ESTABLISH PROBABILITIES

The aforementioned answer-variable distribution is nothing more than an array of x-axis values — usually sorted from smallest to largest (see Chapter 10). A risk assessment would not be a risk assessment without associating each x-axis value with its appropriate probability (see Output from Monte Carlo Analysis — The Frequency and Cumulatve Frequency Plots section of Chapter 9).

Each point plotted on a risk-model-output curve contains two pieces of information. One is the magnitude of the coefficient. For example, on a plot of profits, each point plotted would represent the number of dollars we could realize as profit. In addition, as part of a cumulative frequency plot, each point is assigned a probability that indicates the likelihood that you could attain that value or a value of greater magnitude.

Probabilities such as those described here, expressed as a cumulative frequency plot, aid significantly in the decision-making process. For example, if we need to realize a profit of $50MM from a venture and from the cumulative frequency plot we surmise that there is a 95% chance of realizing $50MM or more, it is likely that we will pursue the opportunity. However, if the plot indicates that there exists only a 45% chance of achieving our base-case profit, then the decision to go ahead with the project is not so cut and dry, so to speak. With a 45% chance of economic success, one would have to contemplate the avenues of gathering more information, weighing this opportunity against other opportunities, assessing the company's ability to sustain a loss if the opportunity fails, and other considerations to help us make our decision.

COMMUNICATION WITH CUSTOMERS
AND THE PUBLIC

Risk models and the output from such models are potent vehicles for communicating with customers and the public. This communication avenue can be utilized in every business.

Consider the scenario in which you have been presented with an investment opportunity. The investment will require the outlay, on your part, of $100,000. Without a risk assessment, those attempting to entice you assemble a proposal that explains just how and over what time the return on investment will be generated. The pitch presents you with a deterministic estimate of the investment's return to you. The single value presented is $150,000 — a $50,000 profit. This appeals to you, and you decide to invest.

The investment is made, but because of bad weather, a short growing season, labor problems, and unforeseen transportation snafus, the investment returns only $100,500. You are outraged because any other competing investment would have had a superior return.

The scenario in which a risk assessment is applied is much different. Suppose the investment proponents engineer a risk model to estimate the range of possible returns. The model includes variables that force consideration and quantification of delays, the probability and consequence of labor problems, transportation availability, weather calamities, and so on. All of these parameters are represented as distributions.

The model terminates with a distribution of potential returns. The potential-return curve indicates that if there is a confluence of nearly all potential problems, there is a small but significant chance that nearly the entire initial investment of $100,000 could be lost. The most likely return is around $135,000, with a maximum, but unlikely, payback of almost $175,000.

Well, now you have to consider the distinct possibility that all of the investment could be lost and that there is a significant chance that the investment could return less than the initial $100,00 you might put in. The reasons for the possibility of loss have been explained to you variable by variable. After long consideration, you decide to go ahead with the investment.

Under this scenario, when the investment returns only $100,500, you are disappointed, to be sure, but are actually relieved that the entire investment was not lost. In either scenario (deterministic or stochastic) the result is the same, but your reaction to the final value is drastically transformed by the application of risk assessment. See Chapter 4 for more information concerning risk and communication.

Consider now a lawyer who has used a risk model to calculate the potential damages for which his client may be responsible. Like the investment risk model described above, the legal model yields a range of potential liabilities. In the case of the investment risk assessment, communication centered mainly on the output variable representing return. Although the lawyer certainly would discuss in detail the output plot of expense, he would use the risk model primarily as a communication tool to foster discussion concerning the input variables for the model.

A comprehensive legal risk model typically contains a dozen or more input variables that are combined to yield the resultant liability distribution. Discussion of each of these variables and the range of values used to represent each variable is invaluable communication with the client. When the client understands and agrees to the range of the input values, the resultant distribution of potential liability is more easily accepted and most certainly is better understood.

PROBABILISTIC BRANCHING

It has already been mentioned that an output plot from a risk model conveys information regarding the probability of occurrence of a range of output values. However, probability can also be used to great advantage as an integral component of the risk model/equation logic.

Let's consider that we are in business with a partner. Our partner's business has fallen on hard times and it might go out of business in the coming year. One of the projects in our company depends upon the contribution this partner. We would like to build a risk model to project the profits from our partnership project, but we realize that this risk model must consider the possibility of failure of the partner's business.

Meetings within our own company and with partner representatives lead us to believe that there is a 60% chance that the partner's business will survive the current financial downturn. Therefore, conversely, there is a 40% chance of economic failure for the project.

Our risk model, which calculates the net-present value of the project, will then contain some probabilistic branching thus:

```
SUCCESS = 60
If Rand_num < SUCCESS then
    (execute success plan)
Else
    (execute failure plan)
```

where Rand_num is a randomly generated value between 0 and 100. The lines "execute success plan" and "execute failure plan" represent sets of equations that would calculate profit under each scenario. Our output plot of profit, therefore, would reflect the probability of our partner's failure. This is a simple example of a powerful and oft-used risk assessment technique.

RELATIONSHIPS BETWEEN VARIABLES

The real world is a complex place. Most occurrences in the real world have an effect on other happenings. If the price of food rises, it might delay the purchase of that new car. As demand for your product rises, the price you get for that product also can rise. As labor costs and taxes increase, your profit margin per unit might decrease.

Through the application of dependence (see Chapter 12), a properly constructed risk model can closely emulate real-world events. Let's consider the situation in which we would like to launch a new product in the marketplace. Three (of many) variables in our risk model are stage of technology (SOT), customer commitment (CC), and maximum cash exposure (MCE). From these and other variables, we would like to calculate the first-year net income for the potential new product. We believe that the technology lead on the competition is significant. However, it is not the actual technological edge that counts, but the customer's perception of our technological prowess that is paramount. Therefore, the distribution representing our stage of technology ranges from relatively low values to relatively high values because of uncertainty surrounding public perception.

We also believe that as our SOT is more favorably perceived by the public, our CC will increase. People like to "go with the winner" and prefer to buy from a company they believe will be in existence in the years to come. So, in our risk model, we establish a relationship between SOT and CC so that as SOT increases, so does CC.

As our SOT increases, it is easier and easier to use the money of investors, rather than our own cash, to fund the company. Investors are drawn to companies that are perceived to be leaders in their field. Therefore, as SOT increases, MCE (spending our own funds) decreases.

In this risk model, then, SOT is "controlling" the variables CC and MCE. If, on a given iteration in the Monte Carlo process (see Chapter 9 for more details), a high value of SOT is selected, the dependence mechanism will cause it, in this case, to be combined with a relatively high value for CC and with a relatively low value for MCE. Conversely, a low SOT value will be linked with a relatively low CC coefficient and a relatively high MCE number. Risk models that honor such relationships between variables can do an excellent job of emulating real-world processes.

LINKING OF DISCIPLINES

As alluded to previously, the real world is a complex place. A single opportunity generally is composed of many parts and represents multiple disciplines. A well-engineered risk model can logically and succinctly meld multiple disciplines into a single entity.

Consider the situation in which our environmental firm has been solicited to bid on a remediation project. The job would be to clean up surface and subsurface liquid waste. This will be an intricate, delicate, and costly venture. Our firm, of course, would only bid on the project if we thought we could profit from the effort.

To estimate the potential for profit from this project, we call together the heads of all departments. Our political analyst attends the meeting because we know from past experience that we likely will encounter political resistance in some shape or form. In the past, local opposition groups halted projects because they contended that proposed clean-up efforts fell short of what actually was required. We have also run into trouble with conflicting state, county, and local regulations. In addition, required permits often are slow in being issued at the state capital because of political wrangling.

Our scientists also attend the meeting because this project presents scientific challenges. We will have to be able to benchmark the level of contamination of the current situation and then monitor the toxin levels as we proceed with the cleanup. The offending substances can be toxic even in exceedingly minute quantities. Our scientists are not sure that we can make statistically substantiated claims that we have reduced the toxin concentration to the required level, given the low levels of concentration.

The meeting includes representatives from our engineering staff because there are myriad engineering challenges. Building special sampling equipment will be an especially daunting task. Also, we will need to drill many wells to monitor the subsurface toxins. We have to do this without disturbing the water table and without causing more surface pollution by the drilling process itself. Other engineering challenges include transportation and disposal of the toxic waste.

Representatives from our commercial arm attend the meeting because commercial aspects of the project also need to be considered. The cleanup effort will require the purchase of large quantities of special chemicals. There are only two companies that produce the desired commodities, and it is not clear whether we can obtain the necessary volumes in the time allotted.

Freight rates are astronomical for the transport of toxic waste by rail. The cleanup site is quite distant from the nearest rail line, so trucks also will need to be purchased for initial transport from the site.

Logistic, environmental, financial, and other considerations also are concerns. A risk model is a unique entity by which all aspects of a problem can be joined to result in a desired outcome (see the Comprehensive Risk Assessment Example section in Chapter 14 for details and examples). Prior to the use of risk assessment technology, such problems would be approached quite differently. Each department, usually without coordinating with other departments, would put forth their deterministic estimate. These individual values, then, would be aggregated to arrive at a single resultant value.

Results reached in such a manner rarely come close to the actual real-world value. Ignorance of the dynamics and interactions of the individual departments was alone enough to add significant error to the estimates. A well-constructed risk model, through the application of probability modeling, dependence, and other techniques, generally alleviates this malady. In addition, the iterative nature of Monte Carlo risk models attempts to ensure coverage of all reasonable permutations. Risk models, therefore, are a powerful and convenient means of bringing together and correctly integrating data from multiple disciplines.

CHECKLIST — DON'T TAKE IT FOR GRANTED

One desirable attribute of a risk model is that it affords a list of variables. A user of a risk model is, in one respect, like an airline pilot. A pilot may have flown the London-to-Newcastle route 1000 times in the past. However, each time she climbs into the cockpit, she reviews the checklist of items that must be inspected or tested prior to the flight. So it is with risk-model users.

Consistent application of a risk model forces individuals to consider each and every element of an analysis and to not take anything for granted — no matter how many times the individual has previously executed the model. Many is the time I have seen that the sources of error in an analysis lie hidden in the items that were "assumed" or taken for granted or as givens.

Utilizing risk models forces not only acknowledgment of each variable, but fosters discussion concerning the range of values, etc., that will be used to represent that parameter. Even the most mundane and "assumed" items receive this treatment. This can be a critical practice.

For example, in the oil-well-drilling business there are parts of the world where there exists plenty of what is called "source rock." These are the rocks that contain ancient plant and animal matter and, when deeply buried, produce oil and gas from the organic material. The oil and gas produced in these rocks migrates upward through the rocks of the earth's crust until, hopefully, it is captured and held by any number of geological mechanisms. We call these places of capture "traps."

An oil-and-gas risk assessment model typically contains subsets. Each subset of the model addresses some aspect of the petroleum system. Just a few of these aspects are the richness of the source rock, its volume, depth of burial, geothermal gradient, migration pathways, trap-sealing mechanism, porosity, permeability, water saturation, trap volume, recovery factor, and formation volume factor.

In the special areas of the world where it was traditionally assumed that shortage of source rock was not a problem, users of an oil-prospect-risk-assessment model typically "turn off" the source-rock section of the model. They would do this by supplying default coefficients that would generate far more oil than could be held in any single trap. When problems arose regarding matching oil reserves from less-than-half-full, real-world traps with the reserves estimated by the risk model, the explorationists, for a while, were stumped. The problem, as it turned out, lay in the "assumed" variables. In fact the traps being tested were not originally filled with oil because of a local lack of source rock and deficient migration pathways. The lesson learned in this case was to use the list of risk-model variables as a checklist and to consider each variable for each execution of the model. Risk models, therefore, serve as an excellent means of enforcing completeness and consistency in any evaluation process.

UPDATING ESTIMATES

The world is a dynamic place. Things change all the time. As soon as you complete an exhaustive and comprehensive evaluation of a situation, it is out of date and practically useless. Such is life.

Risk models afford us the means to at least attempt to keep pace with the ever-changing world. A properly designed risk model is not just a set of probabilistic equations, but is a user interface by which users can nearly effortlessly and without having to supply statistical parameters (mean, standard deviation, etc.) enter data into the system and cause it to execute. A fully integrated risk model also should serve as but one link in a process chain. That is, a complete evaluation process should involve a data gathering and verification step (see Users Will "Game" the System — The "Risk Police" section in Chapter 3), a data processing function that perhaps includes a risk assessment, and a portfolio-management function. In such comprehensive processes, it is not uncommon for the user interface of the risk model to serve as the primary user interface for the entire system. That is, all data are either passed to the risk model from data sources such as databases, or are entered directly into the risk-model user interface.

In such a configuration, a user can approach the risk-model interface and "call up" the previous data entered into the risk model. If the changing world has caused a portion of that data to become obsolete, a user can utilize the risk-model interface to quickly and easily supplant the obsolete entries with relevant data and re-execute the entire process. This process can, in a properly designed system, result in refreshed risk-model output and a new mix of opportunities resulting from the linked portfolio-management process.

In this way and others, an integrated risk assessment system can afford a quick and nearly painless means for updating databases and for making business decisions. The real power of a risk model only is realized when opportunities of similar ilk are consistently assessed and when the risk model is executed as part of a process that results in portfolio management.

As an example of this process, let's consider a brokerage house that maintains a portfolio of investments for clients. This portfolio is designed to generate the highest return on investment at a given risk level. The portfolio is dynamic because market conditions, portfolio-member companies, and myriad other things change on a minute-by-minute basis.

Firm executives have decided that they would like the ability to update the portfolio on a daily basis. Each day, data are automatically collected and stored in the firm's database. In addition, the firm's financial analysts execute other systems and use their market knowledge and experience to daily generate forecasts, trends, and other information.

Each day, an individual at the firm can approach the risk-model user interface and, using the mouse, double click on a risk-system icon that causes the risk model to be populated with the latest information stored in the firm's database. The interface also affords the user a quick and easy means to enter other forecast and analytical data generated by the analysts. When all data have been entered, a click on another icon executes the risk model. Probabilistic time-series and other stochastic data automatically are passed from the risk model to the portfolio-management routine that uses portfolio-management technology to automatically calculate a new mix of investments. With a properly designed risk model, this process is economically and technically achievable. This process can be a powerful adjunct to almost any business.

TO WHAT TYPES OF PROBLEMS CAN RISK ASSESSMENT BE APPLIED?

Risk assessment is a universal and generic tool. Typical model types include health and safety (fate/transport-type), environmental, transportation, construction, exploration, legal, insurance, and many other types. To impart to the reader a sense of the breadth of application of this tool, I have listed below summaries of just a very few of the models upon which I have worked.

> *Legal model* — A model that calculated the net benefit of settlement vs. litigation was built to aid in legal decisions. Net benefit of settlement, net cost of litigation, net cost of settlement, total cost of verdict, and other outputs were calculated. Input variables that comprised broad categories included parameters such as litigation costs, total damages, likelihood of a verdict, and other probabilistic and economic aspects.
>
> *Environmental Health and Safety model* — Models were built that calculated the total environmental, health, and safety risk and cost associated with entry by the company into various countries around the world. Risk of subcomponents of the model also were calculated and presented. Input-variable categories (many variables per category) included public perception considerations, government approvals/permits, ecological/cultural parameters, health and safety considerations, and evaluation of preexisting damage.

Pipeline Route-Selection model — A comprehensive time-series model was constructed to help a consortium decide which of several routes would be selected to construct a pipeline for a major oil field. The pipeline route-selection model calculated tariffs and other parameters from variables that represented major categories of consideration. These included political concerns, environmental problems, commercial considerations, financial parameters, technical considerations, taxes (for many countries), and other parameters. This model was used with great success to rank and prioritize routes.

Political models — Models were constructed that evaluated other countries based on categories of variables. Categories included political stability, foreign investment conditions, operating environment, transportation infrastructure, and other considerations. Results from the models facilitated comparison of countries on common scale.

Capital Project Ranking and Portfolio Management model — This model calculated profitability index (PI), internal rate of return (IRR), net present value (NPV), and other financial outputs. Input variables included project safety and environmental aspects, cost estimates, incentives, discount rates, taxes, maintenance and insurance costs, and other considerations. This model was run on all capital projects at a manufacturing facility. The projects were ranked and portfolio-managed based upon the model outputs.

New Products model — Research and development organizations generate products and processes, each of which may have commercial value. It is, however, expensive to launch a new product in the marketplace. Therefore, products and processes need to be ranked. A model was built that included categories of variables. Some of the categories were technical considerations, marketing aspects, financial/commercial facets, and others. The model facilitated the application of weights to the various considerations. Results from the model allowed the prioritization of potential new products.

Fate/Transport model — A model was constructed to calculate inhalation exposure. Exposure was represented in the model as the average daily dose for noncarcinogens and the lifetime average daily dose for carcinogens. Among the input variables were parameters such as concentration of chemicals in the air, inhalation rate, bioavailability, exposure duration, exposure frequency, body weight, average lifetime, and other considerations.

2 Risk Assessment — A Process

CONTENTS

Risk Assessment — The Process 11
The Time It Takes to Do It Right and When to Get Started 14
Time for Preparation of the Organization — Different People See It
Differently 14
What to Provide 16

RISK ASSESSMENT — THE PROCESS

The marketplace abounds with risk-assessment software and consultants who are "experts" at applying risk-related systems to the problems of their clients. Such applications tend to be relatively sedulous academic and technical exercises. In addition, these applications generally are aimed at the assessment of risk associated with a single "project" or a small number of opportunities.

In the nearly 20 years I have been involved in statistical analysis and risk assessment, if there is one thing I have learned (and it took me quite a number of years to learn it because, in the beginning of my career, I knew everything), it is that risk assessment is a process. The technical aspects of software selection and risk-model building are relatively trivial when compared to the ergonomic, organizational, political, communication, and educational aspects of implementing risk assessment in an organization. Technical brilliance alone does not come close to being sufficient.

To begin with, risk assessments rarely yield maximum benefit when applied in "one-off" or single-project situations. Risk assessment and risk models should be viewed as a means of consistently assessing multiple opportunities or projects with the aim of comparing, ranking, and/or portfolio managing the independently assessed entities.

The risk process begins with two fundamental elements — a need (usually ill-defined) on the part of an individual or group and a vision held by the person or cadre of bodies that will implement the stochastic solution. A call for help usually is the catalyst.

As detailed in subsequent chapters, the facilitator of the risk process, upon answering the call for help, must cajole, prod, and sometimes trick those in need into realizing and expressing just what really is the problem. This is a nebulous and contentious undertaking, even (sometimes especially) when those in need are doing their utmost to cooperate. If you are contemplating the possibility of acting as a risk-process facilitator and are troubled by confrontation and piqued by chaos, then you should stop now!

Shepherding a group through the process of defining the problem and its component parts is not for the faint of heart or for those with a paucity of will. A blend of interpersonal skills, leadership, patience, and technical competence are requisite. It is paramount for the facilitator to ferret-out the fundamental dilemma and to lead the group toward a resolution that is practical and succinct, and that has been reached by consensus.

To be able to enact this chain of events, the facilitator must from the beginning have a vision of where this rambling train should terminate. That is not to imply that the facilitator should be capable of envisioning the "answer" — typically the facilitator knows less than anyone else involved about the troubled business. However, in a competent facilitator's mind it should crystallize early in the process just which parties need to be brought together, which individuals or groups are likely to be and remain detractors, what disciplines will have to be melded (i.e., legal, environmental, financial, etc.), and whether the perceived problem is a "risk" problem at all.

Asking pertinent questions in proper formats, guarding against creating too detailed or too general a model, being cognizant of attempts at "double dipping," (accounting for the same effect more than once), integrating chance of failure, enacting dependence when required, engineering a comprehensive and succinct set of equations, and other tasks are the primary responsibilities of the risk-model facilitator. Although actually important to the overall process of building and implementing a successful risk assessment process, the aforementioned tasks are primarily procedural or mechanical and represent but the proverbial tip of the iceberg.

Organizational consequences have to be considered. Foisting a risk assessment process or model upon an organization will not only change how opportunities or liabilities are assessed, but will significantly alter the way an organization makes critical decisions. As detailed later in this book, such decision-making processes typically lead to selection of "winners" and "losers." Although these two classes might have evolved under any selection scheme, the risk model is an easy and popular source of and target for the angst felt by those who have been "done wrong." Change generally is distasteful. A strong commitment by the organization to the new and different quantitative and stochastic lifestyle is essential.

Other organizational upheavals are necessary as well. If multiple opportunities are to be assessed, ranked, and portfolio-managed, then it is critical that a common and consensus-built risk model be used to achieve consistency in assessment. Following the launch of a risk process and model, it is not long before those using the model "get wise" to what needs to be put in the "front end" to receive a favorable result on the "back end." The only real guard against "gaming the system" is the imposition of a board of review. Members of the board are charged with reviewing all inputs to the system. Organizationally, this is tantamount to creating "Big Brother." Without careful and delicate implementation, this organizational step can consume all the energy of those attempting to promote this necessary aspect of the risk process.

As you might guess, politics also can raise its ugly head. Just wait until a pet project of a powerful political entity is found by the risk process to be lacking relative to other opportunities and, therefore, is passed over for execution. Staunch supporters can change their stripes when they feel betrayed by the process they helped to create. Genuine political support and an organizational means of consoling those whose projects are not selected should be part of any worthy risk-implementation plan.

Instituting methods of consoling those associated with projects not selected is just one cultural adjustment that must accompany an institutionalized risk process. In some corporate cultures, salesmanship and the "good old boys" system are critical ingredients in the recipe for project success. With the advent of an impartial and quantitative risk process (review boards and all), parameters that in the past were ignored or taken for granted now are quantified, scrutinized, and invariably integrated to achieve a consistent result. Ranking of projects can be a much more businesslike and "bottom-line-oriented" affair. This is a drastic cultural change in many organizations.

An equally difficult adjustment can be the newfound ability to "call the baby ugly." In many businesses, individuals have for years put forward proposals that had a high rate of economic failure, or were not as attractive an economic venture as passed-over competing projects. Those individuals who had previously proposed such less-than-optimal projects will, with the advent of a consistent risk process, have to confront the fact that their baby, like many in the past, is pretty ugly. This can be a wrenching cultural adjustment and should not be underestimated.

Still another salient aspect of risk-process implementation is that of education. Obviously, those individuals and groups charged with feeding and executing the risk model need to be educated with regard to the technical and mechanical nuances of the risk model. This type of education is time-consuming, but pedestrian.

The ultimate educational challenge rests in bringing understanding of the output of risk models to those who will (should) use the stochastic data in their decision-making process. Often I have witnessed the successful implementation of a risk process with the "troops," only to have the process die on the vine due to the inability of decision makers to appreciate, understand, and utilize risk-model output. Systems and processes must be put in place not only to generate risk assessments, but to facilitate the use of risk-model output by the organization. A large part of creating this organizational capability is educational in nature.

The preceding paragraphs outline just a few of the components of a well-constructed and -implemented risk process. While it is true enough that not all (or even most) risk processes and models become formally institutionalized in organizations, each process should be approached as though it will be thus incorporated.

It has been my experience to see such models become very successful parts of a relatively small arm of an organization. Competing or simply related organizational entities take note of the acclaim directed toward the much more efficient risk-process-embracing group. Before you know it, hastily constructed and poorly implemented "risk" processes are rampant and, in most instances, do more harm than good. Poorly implemented risk processes likely are worse than none at all.

THE TIME IT TAKES TO DO IT RIGHT
AND WHEN TO GET STARTED

Throughout this book I will be referring to processes and facilitation techniques that form the core of the risk assessment exercise. These processes and techniques sometimes are iterative, time-consuming, and require the repeated participation of individuals and groups deemed critical to the solution of the problem. The belief that risk assessment is a last step is one that is widely held but nonetheless misguided.

As will be described in detail in later chapters, the process of generating a comprehensive risk model should spearhead any effort. The risk-model process outlined will

- Cause the critical parties to meet;
- Foster discussion of the problem and its components;
- Focus the entire effort on just what is the problem and just what is required as a solution;
- Provide a basis for categorizing the parts of the problem and identifying critical variables;
- Promote discussion of the various aspects of the problem and how they will be represented and combined to arrive at the desired solution;
- Provide a means to delegate and distribute problem-related tasks;
- In some cases, short-circuit the entire effort if it is realized that perhaps it should not be done.

The process of preparing a valid risk model, therefore, should be one of the first tasks undertaken in a project. This task requires the participation of people — sometimes many people. In turn, this requires the skills and talents of a practiced group/process facilitator. Leading a group through the risk-model-building process is, as the old saying goes, a bit like "herding cats." Risk assessments can, and usually do, involve many aspects of a business including financial, commercial, technical, environmental, political, and others. Participants from each of these disciplines (and perhaps many more) may be necessary to the solution of the problem. Opinions on what is important, expectations for the result, methodologies of approach to problems, patience levels, "turf" considerations, and many other human-related aspects need to be considered and deftly handled.

Working with a variety of participants and working through the process toward the solution of the problem takes time. A correctly designed and handled risk assessment process is a thing of beauty. From the start of a project, a well-orchestrated process can focus the effort and participants on a problem that enjoys consensus and on a solution that is clearly defined. Responsibilities of individual groups also should be identified.

TIME FOR PREPARATION OF THE ORGANIZATION —
DIFFERENT PEOPLE SEE IT DIFFERENTLY

Risk models sometimes are designed for and executed on single opportunities in a "one-off" mode. More commonly, risk assessments are performed with the goal of comparison in mind. That is, we generally assess not just one opportunity, but several with the aim of comparing these opportunities so that we might manage our portfolio of options. When this process of comparing and culling opportunities takes place in an organization, it necessarily begets winners and losers.

Most organizations need time to adapt to this mindset for several reasons. First is the aforementioned generation of winners and losers. Typically, prior to implementation of a risk model and the resulting comparison of opportunities (allow me here to call these opportunities "projects"), each project might have been prepared and carried out in relative ignorance of other seemingly unrelated projects. Each project, then, gained support and funding based on its individual merits. With the advent of risk assessment and portfolio management, everything changes. The new process necessarily ranks projects relative to one another on some common ground, for example, net present value (NPV) to the corporation. Under the new risk/portfolio management scheme, projects that were, in the past, considered to be unrelated are being compared to one another. Funding is being given to some and denied to others.

This section of the book is, for good reason, in part entitled "Different People See It Differently." The risk assessment and portfolio management scheme will appear to be a great boon to the corporation. For example, those who sit in positions that make them responsible to stockholders will see this process and the resulting increase in efficiency and revenue as a great step forward. Not so from the perspective of those whose projects are being risk assessed and portfolio managed. From their perspective, this process takes valuable time, often brings to the fore uncomfortable feelings and situations, causes them to quantify and document things that were heretofore "taken for granted" or assumed, and results in projects not being funded. The project upon which they were working might be "passed over" for another project of which they may never have heard.

Under a more conventional project selection and funding scheme, most business units (groups) could be assured of some level of project funding in a given year. With the advent of the risk ranking/portfolio management process, funding is no longer likely to be more or less distributed over all projects. Some will be selected for funding and some will not. Even those project proponents whose projects are selected are likely to resent and/or resist the implementation of a risk process.

Prior to this new process, projects that were funded were not necessarily compared with other dissimilar opportunities. Implementation of the new risk-related process not only compares a project to a wider range of competing projects, but the new process forces project proponents to comply with the quantification of project parameters and other rigors of the imposed risk process. Therefore, even those groups or individuals whose projects are selected for funding feel as though "Big Brother" is watching and that they are no longer in control (if they ever were) of the situation. In addition, funding levels are likely to be similar to funding levels in the past, but those who get funding have to go through much more work than they had previously (for the same result).

So, from the corporate management point of view, this looks great. From the point of view of the people associated with the project, this looks like a process that is a lot of work and results in a significant chance of not being supported. It turns out that no amount of talking on your part (the instigator of the risk process) is going to convince those who have been hurt by the process, or those who could potentially be hurt, that this process is for the greater good and is going to be helpful to them in the long run. Time and continuous interaction with these people are the only remedies for this malady. Only when they are convinced (and this has to be true) that portfolio management of opportunities will result in a stronger company and that the risk assessment process really is a benefit to them, will they "buy into" the process and cease resisting the change. Corporate culture changes take time. Be sure to allow plenty of it if you are likely to head down the risk assessment road.

WHAT TO PROVIDE

The old adage "the customer is always right" should be tempered with "well, sometimes." When it comes to performing a risk analysis, there is no question that the results of the effort should be a model that brings joy and contentment to the hearts of the customers, but rarely is it true that the customers have in mind a practical approach.

The quintessential example of this is nearly any risk model built for engineers. Let me preface this by saying that some of my best friends are engineers! However, when collaborating on a risk model with engineers, it always seems that there is no detail too minute to warrant exclusion from the model. Engineers are not alone in displaying this characteristic. Most folks who solicit help in building a stochastic model are much "too close to the problem" and believe that the veritable cornucopia of complications with which they daily contend are relevant. If you in fact built the risk model they asked for, it would undoubtedly, even for them, prove too large and unwieldy to be practically applied.

In subsequent chapters of this book are detailed the processes that should be applied to ferret out just what is the conundrum to be solved and precisely the methodology that should be used to solve it. I will here preempt the following chapters a bit by stating that the practiced risk-model builder should apply good listening skills and powers of reason and deduction in an effort to state, as simply as possible, just what problem it is that practically can be tackled.

No doubt, details are important, and are purported to be where the devil resides. However, the adages "time is of the essence" and "simple is always best" should be guiding lights. It should always be the aim to build a model that is as "high level" as is practical. For example, you would not build a construction model in which is generated a separate variable for each individual rivet, piece of steel, length of pipe, and so on. On the other hand, it likely is an oversimplification to simply concoct a single variable called "costs." It is a tricky business to build a high-level model that contains sufficient detail to make people happy. High-level models have the advantage of being simpler to "feed," simpler to run, simpler to maintain, and more universally applicable.

Usually, though, there comes a time when every high-level model is modified to handle a changing situation. There are two basic philosophies to the art of model expansion. One is to add on to the model to accommodate the new situations. The other is to raise the model to an even higher level.

For example, we may have a model that evaluates blue cars and green cars. Next thing you know, someone comes along with a red car and says: "Hey, your model can't handle this." Well what do you do? Using the "make it bigger" philosophy, you would add a new "red car" variable to the model. Alternatively, using the "make it higher" principle, you might discard the original "blue car" and "green car" variables for two new variables called "car" and "color." Now you can handle any car of any color without resorting to expansion of the model.

Next, however, someone comes along with a light truck and points out that your model is again deficient. The "make it bigger" folks add a "light truck" variable to their "green car," "blue car," and "red car" model (where does this end?). Many spreadsheets (you know the ones I mean) grow like this over time. The "make it higher" people again attempt to produce an all-encompassing model by replacing the "car" variable with the variable "four-wheeled, gas-powered conveyances weighing less than 4000 pounds." So they now have a model composed of this new variable and "color" which handles all the situations they have seen thus far.

Obviously (or at least it should be obvious), both these model-expansion philosophies can be carried to ridiculous extreme. The model builder's task is to balance these two philosophies so that the model can indeed handle a wide variety of situations and yet be simple enough so that it can be practically applied. Again, subsequent chapters outline the details of this process.

There still are other situations in which customer demands have to be taken with the proverbial grain of salt. It happens, figuratively speaking, that you will get a call for help from the guy responsible for arranging deck chairs on the sinking ship. He believes he has a real problem with the chairs because they keep sliding down the deck of ever-increasing pitch. You, however, taking the broader view, astutely deduce that the problem actually lies elsewhere. This somewhat macabre scenario has its real-life corollary.

Consider the situation in which you have been called to help the Waterfront and Storage Department become more efficient and save some money. Their perceived problem is that they would like to increase dock space for ships and increase on-shore storage. They believe this is necessary because on-shore storage tanks are full, so the ship that delivers raw materials can't be unloaded. The ship, therefore, hogs valuable dock space for an inordinate amount of time. The individuals that oversee the dock and on-shore storage believe they would like you to aid them in the construction of a risk model that will weigh the relative merits of several competing dock- and storage-tank-expansion plans. Because of your past experience with similar projects, you immediately deduce that this is not a construction or expansion problem, but is one of optimization.

The ship is large and, therefore, can deliver vast quantities of raw materials. The ship is fast and can make a round trip from the raw-materials-production site to the dock in question much more quickly than the raw materials can be consumed by the manufacturing process. The ship and crew sit idle for inordinate lengths of time because they can't unload the cargo. Although the dock manager who requested your services takes the parochial view that this is a dock/storage problem, you can see that the source of the dilemma is ineptitude in shipping optimization. The guy who hired you has "real" and immediate problems and does not wish to discuss the relative merits of efficient shipping schedules. You deem it an incredible waste of time and money to go ahead with the very expensive dock-and-storage-expansion plan. What do you do?

Given that the title of this section is "What to Provide," the first thing to do is *not* to provide the dock manager with the model he requested. Prior to communicating your refusal, however, you should put together at least a rudimentary model that, in your opinion, could solve the problem.

It clearly is a waste of potentially productive dock space to have it occupied by a ship that can't unload. Even worse is the underutilization of the ship itself and the payments to crew members who are, essentially, idle.

The ship is too fast and of too great a cargo capacity for the job. You might hypothesize the use of two barges in place of the ship. Compared to the ship, barges are cheaper to buy, operate, maintain, and crew. Barges also can be of significantly smaller capacity and are slower. If sailing schedules for the barges are efficiently synchronized, the dock-manager's problem would be nonexistent. With the use of smaller, slower barges, everybody wins. No funds are used to modify the docks or on-shore storage facility. The dock space is used only long enough to unload the barges. Barges can always unload immediately because they carry less cargo and are correctly scheduled. The expense of the ship and crew is, relative to the cost and crewing of barges, greatly reduced.

A risk model should be constructed to emulate the barge scenario and to emphasize the aforementioned advantages. When risk-model results are presented to the dock manager, it should be in a meeting that includes a representative from the shipping department who is empowered to recommend action on the barge scenario. If a shipping representative is not present when the potential solution is presented to the dock manager, the dock manager is likely to reject the proposal. His rejection is probable because the proposal recommends solutions that he has no power to enact (under the proposed plan, taking the "heat" off him depends on someone else doing something) and because it does not, in his view, solve his immediate problem. If a shipping representative buys into the plan, then clearly the dock-expansion proposal is a waste of money and effort, and the dock manager is relieved.

So, what is the point of this maritime diatribe? The point is, don't always be quick to provide just what the customer asks for. I consider it my job not only to provide risk-related solutions to problems, but to also provide the best solution. It is often the case that those seeking help are much too close to the problem, so to speak, and the problem they present either is not the real problem or it is only a small part of a much more extensive malady. Enacting this approach, as with the hypothetical dock manager, is an interpersonal-skills and organizational problem first, and a risk problem second.

At least one more fundamental scenario exists in which honoring a customer's request for a risk assessment and model would constitute providing the requester with a less-than-optimal response. Because application of risk technology and processes can yield very positive consequences for those who embrace them, entities who observe the success of others can misinterpret the application of risk techniques as a problem-solving panacea. It occasionally happens that a risk assessment is requested to resolve an issue that could not possibly be mitigated by application of risk-related technology.

Consider the situation in which you have been called upon to apply risk problem-solving techniques to purchase-order processing. The requesting group is dismayed at the organization-wide time and expense associated with the processing of purchase orders. They are considering the purchase prices and implementation costs associated with several competing vendor-supplied order-processing systems. The selected system would be installed at the order-processing headquarters. Your job, as they see it, would be to build a risk model that would compare the long-term benefits (cash flow over time and NPV) of the competing vendor systems so they can base their selection on a quantitative and probabilistic comparison.

You have worked in various parts of the company in which purchase orders originate. It is clear to you, after preliminary discussions with the group, that the real time-related problems and expenses associated with order processing can't be solved by the implementation of a slick order-processing system at headquarters. Rather, the real wrenches in the order-processing works are organizational in nature, having to do with purchase-location approval chains and purchase-location/head-quarters interactions.

Certainly it would be perceived as presumptuous, arrogant, and a bit insulting if you suggest that they are barking up the wrong tree altogether and that a risk assessment model really would be of little or no benefit. The task at hand, then, is to cause the group to come to the realization that the remedy for the real problem is not a risk model. Well, how do you do that?

Elsewhere in this book I describe in detail the process by which a group comes to consensus concerning just what is the question to be addressed. As the risk-model facilitator, you must lead the group through the process that includes steps to

- Brainstorm and record possible statements of the problem;
- Use an iterative voting process to sort through the brainstorming items and select the one(s) upon which a risk model would be focused;
- Build a contributing-factor plot for the problem on which the answer and all contributing factors are plotted.

As a stealthy and wily facilitator, you should in the brainstorming and/or sorting sessions be able to guide the group to state just what is the actual problem. They may believe at the session inception that the question to be answered has to do with comparing long-term economics of competing vendor systems. However, you should challenge the group to examine the problem in greater depth and guide them toward the "mutual" revelation that the problem they originally asked to be solved is not the problem at all.

If the process gets as far as beginning to build the contributing-factor diagram and they still have not realized just what is the problem, you might have to delicately suggest it yourself (the result of a sudden flash of inspiration, of course). In this case, you should make every effort to get them to understand just what is the actual problem. You should not provide them with the risk assessment or model that was the focus of their original request.

3   Consistency — The Key to Risk Assessment

CONTENTS

Biggest "Bang" from Assessing Multiple Opportunities 21
"Type" Consensus Models 22
Common Measure 25
Risk Assessment Vision 26
Sometimes Better to Be Consistent Than to Be Right 28
Locking the Model 31
Users Will "Game" the System — The "Risk Police" 32
Databasing Inputs and Outputs 33

BIGGEST "BANG" FROM ASSESSING MULTIPLE OPPORTUNITIES

It is true that there is benefit in assessing single and unique projects or opportunities with unique risk models. However, the maximum return on invested time and effort only is realized when multiple opportunities or projects are consistently assessed for the purpose of comparing, ranking, and portfolio-managing the assessed entities.

Within any organization, it is not strange to find that n number of projects of similar type are evaluated by n number of unique risk assessment processes and models. Typically, the assessment models are huge spreadsheets that have been handed down "from generation to generation." It is not surprising to discover that the people using such spreadsheets did not write the spreadsheets and have only a cursory familiarity with the internal workings of the model. The quintessential user of such a model can "feed" it the information required as input and can capture and (sometimes) use and interpret the output, but cannot in detail delineate the process and calculations that link inputs to outputs.

Even within relatively small and focused business entities, it is not unusual to find that people whose offices are juxtaposed and whose tasks are similar are employing dissimilar methodologies to projects or opportunities of similar ilk. For example, within a group charged with the responsibility of evaluating the economics of plant (factory) construction, it is not atypical to discover Sally in room 101 using a spreadsheet that calculates NPV from a 20-year time series of cash flows while John in room 102 is employing a different NPV-generating spreadsheet that considers 25 years in its calculations. In fact, it is typical to discover that the spreadsheets, although evaluating similar factory-construction projects, do not require as inputs exactly the same information and link input values with output values utilizing different sets of equations.

The unsuspecting manager of Sally and John believes that, when he/she receives the NPV values from the two subordinates, a valid comparison of the two projects can be made. The inconsistency of evaluation practices precludes any such comparison.

A unique assessment model applied to but a single project or opportunity makes good business sense when the project or opportunity to which the model is applied truly is one of a kind. For example, a corporation may be considering a stock split. A "one-off" risk model might be generated to assess the consequences of such a move. Such business maneuvers are not frequent events. The risk model constructed to assess the effect of the split, regardless of its comprehensive nature, may not be suitable for application to other business problems and will be used only once.

"TYPE" CONSENSUS MODELS

If it has been decided that a consensus model (or models) should be built (see Chapter 5), the logical next step is to decide just how many consensus models are required. For example, management of a construction company might decide that they could bid more competitively for plant-construction jobs and have a better chance at making a profit on jobs awarded if they could better estimate the range of construction costs and associated probabilities. Because their company, at any given time, is being considered for multiple projects, they foresee great benefit in consistently assessing the costs and manpower demands of the various projects. In addition, they would like to better predict the probability of being awarded a given project (they could be "killed" by success if they were to be awarded all projects upon which they have bid).

Management of the company might convene a meeting with department heads to discuss the plausibility of building such a model. In that meeting, management might discover that meeting participants believe that many such models might need to be constructed. It could have been the vision of management to build a single comprehensive "plant-construction" model, only to discover that the multitude of plant types may preclude this.

Department heads who work on chemical plants might voice the opinion that chemical plants are sufficiently distinct from other plant types to warrant a separate and unique chemical plant construction model. Waste treatment plant construction department leaders feel similarly about their project types as do the department leaders who oversee power plant construction projects.

The problem, however, is even worse than this. People reporting to the head of chemical plant construction submit that separate models will be needed to evaluate "dry-chemical" plants and "liquid-chemical" plants. Similarly, those reporting to the head of power plant construction contend that the process of building a natural-gas-fired power plant is completely different from that of hydroelectric power plant construction. They argue that separate models are required. Waste treatment plant proponents intimate that building a plant that treats chemical waste is very different from constructing a treatment facility for solid waste.

This dilemma is not unique to the construction industry. An environmental firm might decide that they could better estimate cleanup and remediation costs, and compare and rank potential projects, if they consistently assessed their opportunities. The boss would like the ability to literally lay down on a table the results from individual assessments so that she could rank them relative to one another. However, when she proposes the idea of a comprehensive and consistent fate/transport-type risk model to the staff, the response of the staff is disheartening.

Those staff members responsible for assessing airborne environmental hazards contend that their charge could not possibly be assessed by the same model that is used to evaluate subsurface flow of fluids. In addition, the "airborne" assessors believe that they will need four separate models: one for healthy adults, one for healthy children, one for adults with respiratory problems, and one for children with reduced immunity. Likewise, the fluid-transport-assessment team members believe that they will need at least two models. They contend that separate models for toxic and nontoxic hazards need to be constructed.

When faced with such dilemmas, whether building plants, evaluating environmental hazards, or attempting to risk assess entities in any other enterprise, it is essential that someone be empowered to determine just what will be the purpose of the risk assessment model. In the section "What is the Question? — Most of the Time and Effort" in Chapter 5, I address in detail the process of how to resolve "just what is the question?" so those details will not be reiterated here. However, it bears mentioning here that the managers of both the aforementioned construction and environmental firms need to decide just what it is they really want to know.

Subordinates at both firms argue that what they do is unique and warrants special consideration. This is typical. Management, however, contends that the firms need to do a better job of ranking and prioritizing their projects across the board. Both camps have a valid point.

It is a cogent contention that different types of plants and various sorts of environmental problems warrant specific and distinct risk models only if the objective is to assess in detail each of these seemingly unique entities. However, if the aim of the risk assessment is, as management would like, to create a capability to manage the diverse portfolio opportunities, then a general model that can accomplish this task must be assembled. Well, just how do you go about that?

In Chapters 5 and 6 I delineate the process by which the critical elements of a risk model are defined and assembled. The astute manager will first assemble representatives from groups that purport to require specific risk models to determine just how many truly unique entities actually exist. An extensive knowledge of the business, insight into personalities, and a grip on available resources and time should all be utilized in deciding just how many truly separate situations exist.

To evaluate each unique opportunity type, a distinct "type" model may need to be built. The savvy manager will enlist the aid of a facilitator skilled in guiding groups through the process of building a risk model. This process begins with deciding just what is the question (see Chapter 5) and includes the step of constructing a contributing-factor diagram (see Chapter 6). It should be emphasized here that unique risk models should not be created for each project, but for each type of project. All projects of similar ilk can then be consistently evaluated.

When each group has successfully completed at least a preliminary contributing-factor plot, the manager whose wish it is to compare, rank, and portfolio manage the menagerie of distinct projects should collect and compare the individual contributing-factor diagrams. The purpose of the comparison should be to identify common factors in the diagrams. Because it generally is the details of different project types that make them unique, the individual diagrams typically contain a multitude of components (even if they are labeled by different names on the individual diagrams) that are common to all diagrams.

For example, at the environmental firm, the facilitator may have led the groups to the consensus that they actually required four "type" models. They decide they need one to evaluate the effect of airborne contaminants on healthy humans, another to assess ramifications in humans with some susceptibility to airborne maladies, another to assess subsurface flow and migration associated with toxic substances, and a fourth model to consider remediation associated with nontoxic subsurface substances. Upon inspection of the four contributing-factor diagrams, the manager should notice a host of common elements. She might note that each diagram contains a variable that collects the number of internal man-hours the project will demand. Another common parameter is the cost of external firms. Three more shared elements are the probability of governmental intervention and delay, the length of such a delay, and how much such a delay might cost in terms of dollars per day. Two additional common properties are the total number of days allotted for the project and the number of dollars per day the environmental firm might charge. Still another common element is the beginning date of the project. A list of these and other shared elements should be prepared.

Now, the manager herself has to go through the "just what is the question?" exercise. If the manager would like to compare, rank, and portfolio-manage the projects based upon timing and potential profit associated with each project, then it should be determined by inspection whether the assembled list of common factors afford the capability of comparing projects based on the selected criteria. If it is determined that comparison is not possible, the manager should either ask the individual groups to collect the necessary additional information in each of their models or decide that she can live with a comparison of models based on the available list of common elements. When an acceptable set of common parameters has been assembled, the manager should build a contributing-factor diagram to serve as the basis for the project-comparison model.

Building type models is a salient first step toward consistent risk assessment and meaningful portfolio management; however, it alone does not assure success in risk-process implementation. Several insidious "bugaboos" need to be dealt with, not the least of which is what I call the "counting-the-chickens" problem.

I illustrate the counting-the-chickens problem by envisioning five houses built in a circle. The houses, therefore, share a circular back yard. In the yard are a dozen chickens. We knock on the front door of a house and inquire as to the number of chickens they have. The occupants glance out the back window and report that they have a dozen chickens. Repeating this process at each house leads us to believe that 60 chickens exist.

So what does the quaint chicken story have to do with real-world risk assessment? Consider the construction firm that has distributed a type construction risk model to each of its divisions. Each division is about to propose a year-long construction project. Each project will require many tons of concrete to be delivered over the year. The amount of concrete required by each project will stretch the sole concrete-supply company to the limit. However, each division, in ignorance of other divisions, populates the risk model with values that, taken alone, look reasonable. Unless some organizational mechanism is employed to ensure that the corporation is not expecting to have n times (n = the number of construction divisions) the amount of concrete the sole provider can deliver, each construction plan, and likely the corporation itself, is in jeopardy.

Just one more (of many) form the counting-the-chickens problem takes is that of customer counting. It is not uncommon, in large corporations, to find that far-flung offices around the world that produce similar or identical products have targeted the same populations as primary customers for their product. Again, only organizationally driven coordination efforts can hope to avoid this counting-the-chickens problem.

COMMON MEASURE

The ability to rank, compare, and portfolio-manage risk assessed opportunities is predicated on the assumption that the assessed opportunities can, in fact, be compared. Ranking of opportunities requires that all opportunities be compared on the basis of one or more common measures. Typically, these measures are financial, such as net present value, internal rate of return, and others. However, measures that are not financial also are common. Parameters such as probability of success (technical success, economic success, etc.), product production, political advantage, and others can be used to compare opportunities.

If the goal of the risk assessment program is to assess individual opportunities/projects for the purpose of ranking/prioritizing/portfolio-managing the assessed entities, then consistency in assessment is paramount. Consistency in the comparison of opportunities or projects of similar ilk is achieved by assuring that several things are true.

First, all opportunities/projects of a given type need to be compared using a common measure or set of measurements. For example, risk models that culminate in a financial analysis often generate risk-weighted net-present-value plots, risk-weighted internal-rate-of-return plots, or plots of analogous financial parameters. Most risk models result in the calculation of distributions for several such values which are then algorithmically combined to produce a single representative measure. This resulting parameter can subsequently be used in the portfolio management process.

In addition, each model must consistently generate the common measure. For example, two risk models may culminate in a net-present-value measure; however, this does not mean that the individual NPV values serve as the basis for a valid comparison. For example, one of the most common inconsistencies in calculating NPV is the number of years over which projects and their NPVs are considered. It is not atypical for a given risk model to calculate NPV from a 10-year projection of cash flows while another NPV-generating model considers cash flow over 20 years. Inconsistencies like this and others can make folly of the use of a seemingly common measure.

Although a noble aspiration, attaining a common measure from divergent risk models is not always practical. For example, there may be valid reasons why one risk model should generate an internal-rate-of-return (IRR) measure while it makes more sense for another model to culminate in an NPV calculation. In still other situations, it may make business sense for one risk model to result in a measure such as "tons of product produced" while another model generates a financial measure such as NPV.

In situations such as these, if comparison and portfolio management of these opportunities is paramount, you should attempt to create "normalizing" routines that will convert one or more of the risk-model-output parameters to a common measure. Examples might be algorithms that convert IRR values to NPV (with, perhaps, the collection of some additional information) or a routine that converts "tons of product produced" to cash and, eventually, NPV. The employ of such "normalizing" or conversion routines will be necessary if portfolio management of multiple and diverse opportunities is the goal.

RISK ASSESSMENT VISION

The preceding two sections of this chapter should lead to a vision and process for an organization that will afford it the ability to consistently assess and subsequently portfolio-manage its set of opportunities. Defining a vision of how you think this whole risk effort should work and where it should go is absolutely essential. It is important, when called upon to implement a risk process, to at least appear, after a reasonable time, to have a sense of what route the effort will take and where the work should terminate. Communication of the vision bolsters the leadership position, lends credibility to the plan (and buy-in), and lends focus to the entire undertaking. These are critical elements of successful implementation.

Once again, as the title of this book portends, the vision must be seen as practical. The initial steps of a vision should include steps to which people can relate and that they believe can be achieved. Setting too lofty a goal as an initial step can be discouraging rather than motivating. Too paltry a goal also can damage the credibility of the plan and can be interpreted by implementers as patronizing. Initial goals of the appropriate caliber will motivate people to achieve them and will impart a true sense of accomplishment when the initial goals have been realized.

Communication of this vision is essential. Often, a graphical representation of the plan is helpful, such as the diagram in Figure 3.1. The vision should be as nonthreatening as possible and should include enough detail so that those to whom it is presented can see where they play a role.

Implementation of such plans is bound to cause some people or groups to, eventually, change the way they presently do things. The plan should be presented in such a way that the processes by which they currently do things fit into the initial proposal. People and groups who will have to change their ways should be presented

Ultimate vision - business unit-to-business unit comparison of opportunities

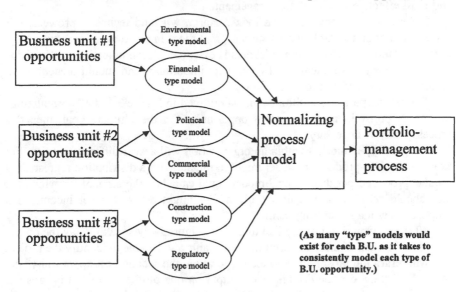

FIGURE 3.1   Risk assessment vision.

with a rational plan spanning a reasonable amount of time so that they can feel as comfortable as possible with the transition.

In Figure 3.1, the ultimate vision is to generate a risk-based system and process by which all opportunities and liabilities of any type within an organization can be assessed, "normalized," and portfolio-managed. Greater efficiency and maximum financial return are the aims. The first step toward this terminal goal is to establish, within subsets of an organization, the likely number of tasks or processes that are repeatedly executed. For example, within a law firm, attorneys may regularly decide which cases can be assigned to "outside" counsel and which must be handled in-house. The same attorneys may repeatedly decide, on a case-by-case basis, which legal actions to settle (and for how much) and which to take to court. Similarly, at a stock brokerage firm, analysts might daily decide which stocks to pursue. They may also regularly decide the proportion of funds that should be directed toward different types (bonds, money market accounts, annuities, etc.) of investments. The applications are, literally, endless.

When at least a few such tasks have been identified, risk facilitators should sit down with the various groups to work through the steps to build a consensus risk model and process (a process explained in detail in subsequent chapters). Care should be taken to ensure that multiple models do not "overlap" in their scope or intent. These models are depicted as ovals at the far left of Figure 3.1.

It is far more advantageous to do a few things right than to exhibit mediocrity at many things. When attempting to implement the "type" models, select a small number of areas that you deem to have the best chance of success. Proliferation of the vision and process throughout the organization, if done right, will follow. The very best way to encourage others to follow the process is to "make heroes" of the individuals or groups that have embraced the plan. Leading by example is the truest and most effective means of encouragement.

In spite of your obvious good looks, charm, wit, and technical prowess, this vision will not go forward without support from those in positions of authority. Buy-in to the plan from organizational leaders is essential. Communication of the vision to such groups or individuals is of paramount importance and should precede communication of the vision to "the troops."

The second part of the ultimate vision outlined in Figure 3.1 is a "normalizing" routine. If portfolio management of opportunities is the ultimate goal, then this process likely is necessary.

Various "type" models within an organization will assess different entities. For example, a plant-production model might terminate with a distribution representing product produced per day. A legal model could calculate the amount for which the firm should settle a case. Still another model might assess the net income of a proposed new manufacturing facility.

What one would ultimately like to do is to, figuratively speaking, lay the output risk plots for each of these opportunities on a table and be able to identify the project that will return the most money to the corporation, the second best opportunity, and so on. Well, how do you do this when output on one plot is expressed in tons per day, another is shown as dollars lost to a settlement, and still another is expressed as dollars of net income per year? The answer is, you don't.

When generating the risk models, every effort should be made to have the models terminate in a common set of measures and units. Typical targets are financial measures such as net present value (NPV), internal rate of return (IRR), and others. It is not, however, always practical or possible to bring groups or individuals around to the realization that it is important that their model share some common characteristics with what seem to them to be completely unrelated risk models.

Portfolio management of diverse opportunities requires a common set of measures upon which those opportunities can be compared and ranked. The normalizing step shown in Figure 3.1 represents algorithmic annealing of various measures. This process can also require collection of additional information from various groups. The result of the process, however, should be a common set of measures (NPV, IRR, or whatever is important to you) that will allow comparison, ranking, and portfolio management of the collection of opportunities.

SOMETIMES BETTER TO BE CONSISTENT THAN TO BE RIGHT

Certainly it is of great importance to be accurate and precise (i.e., "right") when your risk model attempts to predict toxic exposure to humans, estimates project costs, or forecasts any number of other parameters that are sensitive to the magnitude of the output-variable coefficients. There is no question that the aim of such risk models should be to predict as accurately as possible both the magnitude of the calculated variable (x-axis coefficients on a cumulative frequency plot) and the associated probabilities (y-axis coefficients on a cumulative frequency plot). However, there is another class of risk-related problems in which striving for accuracy and precision is of lesser importance and, in fact, may be a completely unmanageable pursuit.

Let's first consider the simple problem of ranking a group of adults by height. My boss has been through the "just what is the question?" process. Because his job is to arrange people on a stage for a choral presentation (tallest people in the center of the back row, etc.), he has decided that he is interested only in rank-ordering the people with respect to their height. He has delegated to me the task of assigning the members of the choir a place on stage according to their height. I realize immediately that I have two problems. One is that I will not see all of the members of the choir together again until the night of the dress rehearsal; however, they will be stopping by individually throughout the day to pick up their costumes. The other problem is that there is no accurate measuring device at the theater, and I have no practical means of obtaining such an instrument before the first members stop by.

At first I resolve that I am in a pickle from which there is no practical escape. After giving it some thought, though, I astutely conclude that I don't really need to accurately determine the heights of the chorus members but only have to know their relative heights for the purpose of ranking them. Having tumbled to this conclusion, I go in search of a measuring device. I stumble upon a wooden stick used to prop open a window. By my estimation, the stick appears to be about 1 foot in length.

Through the day, as the members stop by, I measure their heights using my stick. I quickly realize that the stick is not 1-foot long, because the first person's height (a person not much taller than I am) turns out to be 9.5 "feet" (i.e., sticks) tall. At the end of the day my measurements range from 9.1 to 12.0 "feet," and I know these are not accurate measures of the members' height in feet. However, because I have been consistent in my measurements, having used the same stick on each person, I can exactly and correctly assign a place on the stage, according to height, to each member.

More serious real-world problems can be of the same type: those in which the intent is to rank-order entities. Consider the oil well drilling company that has in its portfolio of prospects 150 wells from around the world that might be drilled. Certainly it will be important to eventually determine which wells will yield the greatest amount of oil or gas. However, the first task of the company might be to determine, using a non-resource-related parameter, which wells have the greatest chance of yielding any resources at all. That is, the company must determine which wells have the highest probability of not being "dry holes" regardless of the amount of oil and gas they might produce. This approach is taken because expensive tests and data collection and processing procedures are necessary to determine the amount of oil or gas a well might produce. Therefore, the company wants to perform those tests only on wells that they deem to have the best chance of being "successful."

While it is true that "success" is partly defined by the amount of resources discovered in a well, the company is first going to rank the wells based upon non-resource-related parameters (i.e., parameters that do not require the expensive resource-related tests and data collection/processing tasks). The company deems this to be a measure of "technical success." The risk model used to determine a technical success value for each well includes a wide range of input parameters. For example, the model collects from the user a distribution representing the range of probability of political success. This is a measure of the likelihood that governments or partners will agree to terms that will allow the company to proceed with development of the well. Another parameter represented in the model by a distribution is the range in probability that the company might run into objections of sufficient strength from environmental groups. These objections could significantly delay or kill the project.

Other input parameters are of a technical nature such as migration efficiency. This distribution-based parameter measures the likelihood that oil or gas successfully migrated in the subsurface from its deep-seated site of generation to the more shallow location of the trap rock into which the company proposes to drill. If preliminary reconnaissance using geological and geophysical techniques indicates that favorable migration pathways are not likely, then the prospect would receive low probability values for this parameter. Likewise, the company may, from preliminary geophysical and geological work, evaluate the probability of the existence of an efficient seal for the trap. That is, it is not enough to have oil or gas migrate into potential trap or reservoir rocks. There must exist a relatively impenetrable (to oil and gas) "cap rock" over the reservoir rock that has prevented, through time, the escape of the oil or gas from the reservoir rock. Based on preliminary geophysical and geological information, risk-model users enter a distribution of values that represent their best estimate for the existence of a sufficient seal for the prospect.

The aforementioned political and technical parameters, along with a host of others, are input to the risk model, which culminates in a distribution of probability-of-success values for each prospect. The methodology for collecting and processing information for each potential well is consistent (or as consistent as reality allows us to be) but not necessarily accurate.

For example, the company may determine that the mean probability of success for prospect A is 30%. Using their consistent process, the company may conclude that the probability of success for prospect B is 50%. Therefore, the company deduces that prospect B is more likely to be technically successful than prospect A and can rank order these and other prospects. However, the 50% value, for example, for prospect B does not necessarily indicate that if the company had in their portfolio 100 identical prospects like prospect B, that half of them would be successful. Calibration of such probabilities to the real world often is a practically impossible task. The company doesn't really know if what they determine to be a 50% probability of success actually is a 25% probability or a 63% probability, etc. For ranking purposes, it does not matter. As long as the company is consistent in their assessment methodology, and as long as they do not use the resultant percentages as terms in other equations (such as multipliers in equations to determine, for example, the magnitude of the potential resource level from the well), the probabilities resulting from the risk model can practically be used to rank order the prospects.

This process of rank ordering entities is common practice in many disciplines (see Qualitative Model and Semi-Quantitative Model examples in Chapter 14). The consistency-in-evaluation approach is a methodology that can make practical the solution to problems that at first seem intractable due to problems related to the collection of accurate and precise input data.

LOCKING THE MODEL

Because consistency in assessment is a salient consideration for risk models that produce output values to be used in portfolio-management processes, preventing model users from changing the model is of critical importance. The only way to ensure that risk-model-output values can be compared and ranked is to ensure that those parameters consistently have been generated.

It has been my experience, especially when utilizing spreadsheets as part of the risk model, that model users will "tinker" with the model parameters or equations to get the model to better fit their individual situation. For example, you might have generated a risk model that calculates the net present value (NPV) for a type of project. Your charge is to rank a multitude of projects from all over the world based on the resultant NPV distributions. In January you distribute your model to the offices around the world, and by March you begin to receive the results from the model, which was executed by users in foreign offices. At first you are happy with the results, but after a critical mass of NPV distributions are delivered, your acute business sense indicates that something does not smell quite right.

You request that the individual offices return the model. You find that each office has somewhat modified the model to better suit its particular situation. Most of the modifications are of sufficient magnitude as to make invalid the comparison of NPVs from different offices. Our ability to rank the opportunities has been destroyed.

To avoid this common dilemma, I have found it essential to work with a risk assessment system that allows the party responsible for model-output comparison to "lock" the model. Locking a model means that the user can input data to the model and execute the model, but does not have the ability to change the input parameters, the number of input parameters, or the equations.

I have found that whenever the issue of locking a model comes up, the groups upon whom the model will be foisted object strenuously because, to them, "locking" translates to "inflexibility" and constraint. A locked model need not be inflexible nor constraining. In Chapter 5, I delineate the process by which a comprehensive, flexible, and consensus risk model should be produced. If this prescribed methodology is followed and skillfully executed, the objections to locking the model should be nil.

I find it useful to be able to lock the risk model at two levels, and the risk system I use facilitates this. One level of locking requires that the person locking the model (generally, the model builder) enter a password. In this locking mode, users can peruse the equations that comprise the risk model but are prevented from modifying the equations unless they enter the password. This locking method works well with large and complex models but seems to fall short when employed with simple models. Astute users of relatively simple models will realize that they can simply copy the equations to a new risk model they compose themselves, modify the equations, and send you the results that you believe were generated by your locked equations, but were not. This behavior necessitates a second level of locking a risk model.

A second level of locking models involves preventing the user from any access to the equations and input/output parameter-building mechanisms of the risk-model-building software. When the model is thus locked, users can only see the "user interface" portion of the model and can only input data, run the model, and review/save/print/etc. the results. Locking models in this fashion can be the only way to ensure consistent assessment.

USERS WILL "GAME" THE SYSTEM — THE "RISK POLICE"

In some situations, in spite of best efforts to lock the model and educate potential users, those executing the model will "game" the system. That is, they will enter values into the system that will, to put it delicately, present their opportunity in the most favorable light.

After implementation of a consistent risk assessment process in an organization, it does not take long for those being subjected to the process to realize just what "hurdle" value the risk model must produce if the assessed project is to have any chance of being selected for execution. In addition, savvy users quickly deduce which input parameters most significantly impact the risk-model results and thus focus on those input parameters so that their opportunity is perceived best. In this situation it is not long before everyone is living in Garrison Keillor's Lake Woebegone, where all the children are purported to be above average. That is, it is not long before all opportunities exceed the minimum acceptable criteria.

The only mechanism of which I know that successfully precludes the input value inflation problem is to establish, for each model type, a cadre of experts that is charged with review of risk-model input. In most organizations this review group is affectionately (sometimes not so affectionately) known as the "risk police." In spite of our society's technical and computer prowess, it is this human review that facilitates the successful implementation of a consistent risk process.

For example, within a company it might be the charge for individual groups to present new and potential locations for outlet stores. Groups whose projects are selected will oversee the construction of the outlet. The company can't afford to build all outlets proposed and thus must rank and prioritize the outlet proposals. This process, necessarily, makes competitors of the outlet-proposal groups.

Company management realizes the competitive nature of the groups and, therefore, the propensity to be seen in the best light. To prevent "inflationary practices" with regard to risk-model-input values, the company has established an impartial review board (the "risk police"). The established risk assessment process in the company requires that before risk-model results for an outlet can be submitted to management for consideration, the input values and resultant distributions must first obtain the approval of the review board. At a risk-model review, it is the job of the review board to

- Review with the outlet-proposing team the values they have used as input to the model.
- Come to a consensus as to the validity of the input values.

- Act as historical reference and organizational memory. This role allows each new proposal to be judged relative to all past proposals.
- Act as gatekeeper. Proposals that are deemed to be in some way unreasonable should not receive the review board's approval.
- Enforce consistency. One of the most critical responsibilities of the review board is to ensure that all opportunities are consistently assessed so that opportunities can be successfully portfolio-managed.

Clearly, established review boards need to be specific to a risk-model type. That is, the company may need to establish separate review boards for outlets that will primarily sell chemicals and for outlets that will market clothing. Regardless of the number of review boards deemed necessary, review boards should embrace the following principles and should exhibit the following qualities:

- Review-board members should be recognized as experts by the organization.
- Review-board members should exhibit impartiality in their assignment.
- Review boards should consist of one or more "core" members for whom assignment to the review board is "permanent" and is not a rotational assignment.
- Review board members should exhibit exemplary interpersonal skills, given that the nature of their assignment is contentious.
- The review board should be recognized as being empowered.

Establishment of such review boards for each risk-model type can be an expensive and time-consuming task. However, establishment of responsible, impartial, and empowered boards of review is the only risk-process technique I can recommend that will result in realistic and consistent risk-model results.

DATABASING INPUTS AND OUTPUTS

A final word concerning consistency will relate to storing the risk-model input values and results. Although, as mentioned in the previous section, the review boards or "risk police" can act as an organizational memory, they can't possibly recall in a risk review all previous input or output values.

If a review-board process has been established, then sanctioned pre-project-execution input and risk-model output values can be stored in a database. The stored values can act as a benchmark for newly evaluated projects. For example, a group proposing a new outlet store might project their sales for a given demographic group to be 10,000 units per month. Values drawn from the database, however, might indicate that sales from similar outlets to similar sections of the population have ranged from 5,000 to 7,500 units per month. The review board might then suggest that unless the group can justify the higher figures, they adjust their sales estimates to match historical values.

It should be noted here the distinction between pre-project-execution values and post-appraisal values. In some businesses it is relatively easy to compare pre-execution risk-model-input values to those that result from execution of the project. For example, a company that estimates convenience-store income for a given geographic location can compare the pre-execution estimates with the actual sales from the store after it is up and running. Both pre- and post-execution values should be stored.

In other businesses, only pre-execution values are generally available. This situation is exemplified by the oil well drilling industry. Companies conducting such business can collect estimates for the probability of a sufficient seal, for a favorable migration pathway for the oil or gas, and for many other parameters. When the well is finally drilled, it either produces economic quantities of oil/gas or it does not. If it does produce sufficient resources, production of the well begins immediately (delayed production is revenue lost) and proponents of the project rarely are afforded the opportunity to verify their pre-drill estimates. If the well is not successful, project proponents are rarely given more funds with which to investigate why the opportunity failed (seen as pouring good money after bad). Therefore, in this industry and others like it, databases contain (primarily) only pre-execution estimates.

4 Communication

CONTENTS

Communication — Just One Shot...35
Communication — Why Bother?..36
Know Your Clients...36
Focus on Impact and Mitigation 37
Diplomatically Handle "Winners" and "Losers" 38
Communicating the Value of Risk Assessment 38

COMMUNICATION — JUST ONE SHOT

Classes certainly play a salient role in the communication process; however, classes only serve as a way to "preach to the choir." That is, lectures to a class are lectures to those who either already believe or who want to believe. A large part of risk communication takes place between the purveyors of risk technology and those who are skeptical, downright hostile, or those whose positions are buffeted by the results of the risk assessment. Excellent communication skills and processes are crucial.

While in the throes of assessing a situation's risk, you may well have the luxury of time — enough time to schedule critical meetings, enough time to make decisions, enough time to gather data, enough time to build and test the model, and enough time to interpret the results. Relative to these time lines, the risk-communication task is a point source. It has many times been my experience that the results of months or years of effort had to be communicated in a single meeting — usually in an hour or less. This is typical in the business world.

Even if you are given as much time as you need in a meeting, it is a rare instance indeed that you are given more than one meeting in which to convey the risk-related information. An example is dealing with the public on environmental issues. After the first "town hall" meeting, local organizations, environmental action committees, the press, your competition, and others are off and running with what you said — or what they think you said! Second chances are scarce, and often ineffective if they do occur.

Given that risk communication must take place in an arena like the aforementioned, it is essential that the person or persons doing the communicating be believable, calm and even-tempered, credible, a good speaker, a better listener, authoritative, and empowered. To be sure, this is a lot to ask of an individual, but it is absolutely essential that the person doing the communicating be capable. Such communication is 50% "the message" and 50% "the messenger." A great message and years of effort can be lost if the message is poorly presented.

COMMUNICATION — WHY BOTHER?

Well, if risk communication is itself so risky, why bother? The communication of risk should be undertaken in almost all circumstances. The most important reason is understanding. For example, in the good old days, lawyers would use terms like "pretty good chance" and "better than average" to describe to clients their chance of a favorable outcome or verdict. Using risk technology to assess legal situations, lawyers can now discuss with clients the range of probability associated with each aspect of a litigation and can quantify the probability of a favorable outcome and the impact of that result. In such situations it is not the quantitative answer that is so important; rather, it is the fact that the risk model parameters and input values become the centerpiece for dialogue with clients. Clients, for the first time, can clearly see what goes into making the decisions and can have meaningful dialogue on the various aspects (see Chapter 14, Legal Model example).

Another example of the benefit of risk communication is that of dealing with partners or investors. Without a stochastic assessment, your best guess at the cost of a project might be $5MM. When the project is complete, the cost was $6.5MM, and the partners and investors are up in arms because you "lied" to them about the project's cost. Results from a probabilistic analysis might have shown that the project has a small probability of costing as little as $4MM, but also has a small chance of costing as much as $12MM, with a most-likely cost of about $5MM. With this information, investors and partners can, prior to joining the project team, decide whether or not they wish to expose their companies to the possibility, though small, of spending as much as $12MM. If they do decide to invest, they will find the $6.5MM figure well within their "comfort zone" and, actually, a relief.

Of course, there also are times when there is no choice concerning whether to communicate risk information. In many companies, if a project is to be approved, it is required that results from a prescribed risk process be presented. Even though a risk presentation may be mandated, good communication skills still are important.

KNOW YOUR CLIENTS

OK, so I'm convinced that risk communication is important and that I should be good at it. Well, just what should I do to help ensure a good communication effort? Good question!

One thing that significantly aids in risk communication is learning to speak "customerese." That is, do your homework and learn the terms and jargon that are familiar to the customer. Of course, it is impossible to become an engineer, for example, if you have not been trained as one. Likewise, having performed a risk assessment will not transform you into an economist. However, it is possible to get hold of materials relating to the subject matter and to the risk assessment and to glean from them at least some of the important concepts and terms.

Knowing too much actually can be detrimental. For example, if you are not an engineer but walk into a den of engineers acting as if it only took a week or so to get a grip on this engineering thing, the warmth of your acceptance by the engineering community will be tepid. Nobody likes a smartypants! It is far better to not appear to know too much, but to make it obvious that you have made an honest effort to learn something about their line of work.

Just as we avoid talking about Uncle Jim's weight and Aunt Milli's wart when the relatives visit, prior to visiting a client it is always a good idea to familiarize yourself with their "sensitivities." For example, I have often been called upon to perform a risk assessment on an ambulatory corporate corpse that expected the results of the risk analysis to aid in resuscitating the deceased. In such instances it must be succinctly but delicately pointed out that a risk assessment would be of little value. In other cases it becomes obvious that the request for a risk assessment is an attempt to set right something that has gone dreadfully awry.

In the aforementioned situations, and others of the ilk, it is important to discover some means of diplomatically refusing to perform a risk analysis, or, if an assessment is made, to avoid "calling their baby ugly." The risk assessment process often can uncover some rather distasteful and, yes, sometimes unwise business practices and questionable financial practices that may have been going on for quite some time. If I had some universally applicable communication solution to the dilemma of illuminating the problem without embarrassing the clients, I would now be writing that solution! The best advice I can impart is to expect this to happen. Investigate potential clients as thoroughly as is practical so you can be as prepared as possible to handle such situations.

In the risk business as in war, there is no substitute for good intelligence. Each client and situation is unique. As stated at the outset of this communications section, it often is true that results of a complex and lengthy risk analysis are expected to be successfully communicated in a short time in a single meeting.

Given that results need to be communicated correctly the first time, it often is good practice to visit with client representatives well in advance of any presentation solely to ascertain just how information might best be conveyed. Knowing that information should be presented in an industrial or academic motif, knowing that it should avoid or stress technical, financial, or other aspects, or knowing that the presentation should be aimed at a specific individual are all examples of the types of intelligence that can be invaluable. Don't be shy or reticent to solicit explicit information from the client organization. Such investigations should be appreciated by the clients, and the presentation undoubtedly will be better received.

FOCUS ON IMPACT AND MITIGATION

The old adage of "pride cometh before a fall" certainly applies to the realm of risk assessment. The risk process can involve the time-consuming collection of much scientific, engineering, financial, environmental, and other types of data. Individuals who solicit and compile such information are rightfully proud of their areas of expertise. As the coordinator of a risk assessment effort, it is your responsibility to temper the zeal of such individuals and control and mold the presentation.

Most clients are not impressed by an overt infusion of technical details or a blizzard of charts and graphs. Presenters should always be prepared to delineate in reasonable detail just how results were generated. However, a risk presentation should focus on the client-organization impact and mitigation facets of the assessment.

If plots and charts are presented, they should convey clearly the financial impact or other data critical to the client for the purpose of decision making. In addition, if mitigative actions are appropriate, the steps and costs of risk-mitigation efforts should be emphasized.

DIPLOMATICALLY HANDLE "WINNERS" AND "LOSERS"

The risk assessment process generally serves as a prelude to portfolio management. The portfolio management process, for all of its stellar qualities, does bring to the organization the dark side of designating "winners" and "losers."

In the majority of situations, a risk analysis is performed to aid in making a decision. Such decisions typically involve selecting one option or, to wax philosophical, to take one road rather than another. The proponents of the road not taken must be dealt with carefully.

As the ancillary risk assessment person, you may well find yourself in the position of having to explain to an organization the results of the analysis. Results may indicate that some members of the audience are no longer going to be involved in the assessed project. You most certainly will not be empowered to offer a place on the "winning" team to those whose proposals were rejected. This situation can be volatile, and deft and delicate handling is required.

As the presenter of risk assessment results, you are not empowered to offer organizational solace to those who find disfavor with the outcome. The only recourse is to emphasize the logic behind the selections made (and you had better have some) and how the decisions taken best address the needs of the greater organization. This sometimes is a difficult or impossible message to successfully convey. The best communications efforts are comprised of the following points:

- The decision that was made;
- The logic and process leading to the decision;
- The benefit to the overall organization.

A discussion of how other decisions would have been worse should be avoided. Certainly it is not beneficial to get pulled into a discussion as to why and how the things that were not selected are worse than the things that were selected. Diplomatic and even-tempered elucidation of the three points listed above is the best communication solution to this difficult but often-occurring communication dilemma.

COMMUNICATING THE VALUE OF RISK ASSESSMENT

In the business world, if a technology is to survive, the value of that technology must be succinctly communicated. Especially in the business world, it is a particularly vexing task to prove that a technology "adds value" (i.e., makes or saves money). In the academic arena, if practitioners understand, accept, and use a technology, then that technology is deemed a success. In the scientific world, if a technology accomplishes a heretofore impossible or impractical chore, then the new method is proclaimed triumphant. In the business area, however, it is the bottom line that counts.

It is a relatively simple matter to calculate the cost of a given technology: "Let me think, I have four people working on this, plus materials, plus computer charges, plus travel expenses, plus…." It is a much more arduous undertaking to document the (dollar) benefit of a technology. So it is with risk assessment.

Those pressed to document the cost/benefit ratio of a technology such as risk assessment typically attempt to associate some dollar value to use of the technology. Don't even bother to go there! Generally, the technology provider is not the primary technology user. Therefore, the provider has to request from the user what portion of the user's financial success can be ascribed to the use of the technology.

In most modern business environments, technology users, in one way or another, are charged for the "privilege" of using the technology. Therefore, when the technology provider requests to share in the riches of success, the technology user sees that request as akin to the seller of a spoon wanting to take partial credit for the culinary extravaganza created by a chef. "We paid you for the spoon — what else do you want?"

Another time-honored roadblock to associating positive value with technology use is the argument, "If you share in the success, then you must share in the failures." Consider the situation in which a new treatment for a fatal disease is successful in preserving the lives of 30% of the patients upon whom it is used. Well, 30% survive, but 70% die (no matter that they would have died anyway). Should you not share in the cost of the failures too?

Another example of this same phenomenon is evident in the world of exploration oil well drilling in which the failure rate is almost always greater than 50% (i.e., more than 50% of the wildcat wells drilled in the world tend to be economic failures). So, suppose we have developed a new seismic technique that is applied to virtually all prospects, and it raises the economic success rate from 40% to 45% in wildcat drilling. Does the new technology only get credited with some dollar portion of the 5% increase, or does it have to share the dollar burden from all those unsuccessful prospects to which it was applied? This generally is a conundrum that remains unresolved.

I have, however, saved the best for last. No one can know the results from the road not taken. Therefore, if a venture that used the new technology is successful, would it not have been successful anyway? Suppose you urge me to wear a special hat while popping popcorn. I take your advice, and each time I attempt to pop corn, it works. This results in a 100% correlation. Is there really any connection? How do you, as the technology provider, respond when the technology user claims that they would have been successful, or would have made that good decision anyway (even without your stuff)? In businesses that are keenly eyeing the bottom line and in which the technology provider has been "paid" for the use of the technology in the first place, the provider should not be aghast at this tenor of response.

So, if the longevity of a technology depends upon the promoter proving and communicating the value of the technology, just how do you do that? Many years of beating this (apparently dead) horse has led me to draw the conclusion that you have to "prove" worth indirectly — not with dollars. One metric to consider and to communicate to those holding the purse strings is the amount of repeat business you enjoy. The real measure of success is not in getting new customers — someone can always be suckered into trying it once — but is in the number of people or businesses that time and again use the technology you provide. Written testimonials from customers also are helpful.

Yet another technique I employ is to request that users of the technology put the methodology on a list of techniques and processes they used in their venture. I ask them to list the technologies/processes in rank order (no scale) with the perceived most useful item at the top of the list. The technology/process thought to be least beneficial should occupy the bottom position. Lists from various users rarely will have many items in common. If, however, your technology consistently ranks near the top of the individual lists, this fact should be prominently communicated to those who have the power to eliminate "unnecessary" items when budget-cutting time comes around.

If I were you, I would be asking myself why I, the reader, was being subjected to this decidedly melancholy diatribe about the near futility of associating value with technology. The answer is because risk assessment is among those techniques to which these anecdotes apply. Typically, a risk assessment forces quantification, documentation, and rigorous integration of elements of a venture. These elements would otherwise have been evaluated and combined by less arduous, "seat of the pants" methods. To adduce that, without risk assessment, risk users would not have made the decisions they ultimately embraced is a claim that usually can't be substantiated. Therefore, like so many technologies, it generally is clear how much cost, time, and effort were required to implement a risk assessment process, but much less evident is the benefit of invoking risk technology.

Risk processes established in organizations do not have a life force of their own. Without constant care, feeding, and promotion, such processes tend to fall by the wayside if taken for granted — only to be resurrected when it ultimately is realized that what was discarded was an essential element of the business. If risk assessment or an established risk process is to survive or flourish, the implementation of metrics and the communication of the compiled supporting data are crucial.

5 Building a Consensus Model

CONTENTS

What Is the Question? — Most of the Time and Effort 41
Consensus Model...42
Group Dynamics..44
Write It Down..45
Sort It Out..46
Group Dynamics Again 47
Units...47
Overarching Categories 49

WHAT IS THE QUESTION? — MOST OF THE TIME AND EFFORT

It happens almost every time. Someone from a group in the organization will call me or send me e-mail and request help with a risk assessment model or process. When I arrive on the scene, typically there is a conference room full of people to greet me. Folks from all walks of life are gathered — managers, political specialists, economists, engineers, other scientists, environmental specialists, legal counsel, and others. I liken this gathering to the three blindfolded persons and the elephant. If you were to ask these three persons: "Do you have an elephant?" their universal answer would be "Yes, we certainly do." However, if you asked each person to describe the beast, the one holding the tail would say it's like a rope, the one touching the leg would indicate that it was like a tree trunk, and the one holding the ear would relate that it was like a big flat thing.

Initial meetings with groups attempting to build a consensus risk assessment share some characteristics with the three blindfolded individuals. If you ask: "Do we need a risk assessment?" the universal answer likely will be "Yes, we do." However, if you ask: "Who thinks they can state in one sentence (two at the most) just why we are here today?" many hands in the room will go up. Each person has their idea of just what question is to be answered. I can tell you from experience that there will be little commonality to their answers. It is an important function of the facilitator to attempt to meld the diversity of opinions and viewpoints into a risk model — a consensus risk model — that everyone agrees is sufficiently comprehensive, rigorous, and flexible to accomplish the task at hand. This is a tall order.

The very first thing to do when beginning to facilitate such a group is to ask someone (who is not the boss) to write on a "flip chart" or board, in a single sentence, just what they believe the problem to be. When this has been done, ask the rest of the group if they agree with exactly what has been written. If they have no other opinion (which they won't if you let the boss do the first writing), you will be in the presence of a very unusual group. Typically, engineers see the question as having to do with the engineering aspects of the problem. Economists likewise see it as fundamentally an economic argument. And so it goes with each person and each discipline (i.e., what does the elephant look like?). Arriving at a consensus as to just what is the problem (or problems) that is going to be solved is a tricky proposition.

CONSENSUS MODEL

When multiple projects of similar type need to be evaluated (like the aforementioned plant-construction projects), a consensus risk model should be constructed with the aim of applying this model consistently to all projects of that type. For details on how to define and build "type" models, see the related section of Chapter 3 of this book.

The ability to rank, compare, and portfolio-manage risk-assessed opportunities is predicated on the assumption that the assessed opportunities can, in fact, be compared. Comparison of opportunities requires that all opportunities be compared on the basis of one or more common measures. Typically, these measures are financial, such as net present value, internal rate of return, and others. However, measures that are not financial also are common. Measures such as probability of success (technical success, economic success, etc.), product production, political advantage, and others can be used to compare opportunities.

For example, let us consider the situation in which a business manager is charged with the task of deciding which of several competing product lines should be selected for marketing. The product lines may be diverse. One line may be footwear, one hunting gear, one basketball equipment, and so on. The manager may decide to compare these opportunities on the basis of net present value to the company. The challenge, however, is to decide whether to allow the departments for the product lines to construct their own assessment models or to attempt to generate a consensus model.

If the comparison of competing product lines is likely to be a unique or rare event, then the manager might decide to allow each department to devise its own assessment model as long as the result of the analysis is the net present value of that product line. If this is the path taken, it will then be up to the manager to be sure that all critical elements (taxes, labor costs, other expenses, etc.) are considered in each department model. Such an undertaking can be a daunting task. This method of comparison may be most efficient in the case of assessing events that are not likely to occur again.

The manager, however, may decide that he/she will many times be confronted with making such decisions. In this case, it likely is more efficient to attempt to build a consensus model which can be applied to the three product lines presently being considered and to all other product lines that may be considered in the future. A graphical example of these two contending philosophies is shown in Figure 5.1. It should be noted that to build a risk model capable of considering all the intricacies

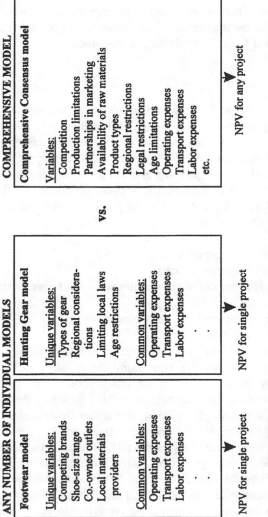

FIGURE 5.1 Example of plans for multiple-individual models vs. a consensus model.

of a multitude of product lines, the manager will have to assemble critical representatives from each department and engage them in the risk-model-building process. This can be a time consuming and sometimes a nearly politically intractable task. However, the effort will pay off with a risk model that can be applied repeatedly to new opportunities or sets of opportunities. Those opportunities can then be compared with options already assessed. Portfolio management is the result.

It should be expected that the risk model will be an evolving entity. With the repeated application of the model to new and slightly different situations, the model will grow and evolve into a more robust and comprehensive assessment tool. Appropriate resources for the maintenance and enhancement of such a model should be considered essential from the beginning of this process.

GROUP DYNAMICS

The real trick in arriving at the consensus solution is to do it without hurting anyone's feelings or having someone feel as though his/her contributions are not being justly considered. There are many approaches used to resolve this type of conundrum; however, they all share the common thread of orchestrating group dynamics.

The first piece of advice I generally give to a group attempting to reach consensus on a question or model is to use a facilitator (discussion/model-building leader) who does not have a stake in the outcome of the project. In fact, the best facilitators are those individuals who

- Have good "stage presence";
- Have good interpersonal skills;
- Understand the process of orchestrating group dynamics in the risk-model-building arena;
- Completely understand risk technologies.

In fact, the best facilitators are those who exhibit the attributes listed above and who have no specific knowledge concerning the problem to be solved. Relative ignorance not only imparts an aura of impartiality, but allows the facilitator to ask "fundamental" questions that otherwise may be ignored or the answers to which would be taken for granted. Discussion of "fundamentals" many times can be of great importance in the process of reaching consensus.

For example, when there is a gathering of individuals concerning a project for which a risk model is to be generated, many of the individuals know one another. This can be both beneficial and detrimental. The benefit is derived from the feeling of comfort that comes with familiarity. This comfort factor, shared by the people in the room, can be used to move things along because introductions are not necessary and "meeting-new-people jitters" is not a factor. Contributions to the discussion are more freely given in familiar surroundings. However, familiarity also can work against you.

In many situations, people who know one another "carry baggage" with regard to past relationships. In addition, there may be competing entities in the room. This may take the form of competing individuals (only one engineer is going to get to

PLANT-EXPANSION MEETING—JUST WHAT IS THE QUESTION?

Individual Statements of Problem

#1 – Engineer 1 - What is the best affordable construction method for the plant-expansion scenario selected?
#2 – Engineer 2 - What is the plant-expansion scenario that yields the highest net present value?
#3 – Economist - Regardless of the plant-expansion scenario (i.e., technical plant-expansion details), what is the net present value of expansion of the product line?
#4 – Environmental Engineer - What effect will spills and other environmental concerns have on the profitability of the project?
#5 – Political Consultant - How will the timing and economics of the project be affected by governmental attitudes and requirements?

Dynamically Kept List of Common Points

Common points between statements 1-2: Scenario and affordable plant expansion
Common points between statements 1-2-3: NPV and expansion
Common points between statements 1-2-3-4: Profitability (i.e., NPV)
Common points between statements 1-2-3-4-5: Economics (i.e., NPV)

FIGURE 5.2 Compilation of ideas for the purpose of determining just what question the risk assessment will attempt to answer.

work on this!), competing groups, or other forms of competition. This and other things such as downright "bad blood" have to be considered when attempting to facilitate a risk-model-building exercise and must be mitigated in a constructive manner. There are many group dynamic processes that can help.

WRITE IT DOWN

After the ice has been broken by the first person expressing their written opinion as to just what is the problem, the facilitator should ask those who take issue with the written statement to write on a piece of paper, in a single sentence, just what they think is the problem. If the facilitator allows verbal expression of the problem(s), chaos results because, invariably, an argument or heated discussion ensues. The facilitation steps to follow next are

- One at a time, have volunteers come to the board and write down their expression of the problem;
- Following each statement, direct the group to find some common elements in previous statements and the newest one;
- Keep a separate written list of the common elements in full view.

In Figure 5.2 is depicted a table that might result from the process described above. In the example shown in Figure 5.2, we are really lucky that a common thread concerning the economics of the project could be found in all points of view. To calculate the net present value (NPV) of the project, we will of course consider the engineering, environmental, economic, political, and other parameters, but at least we now have a common and agreed-upon goal for the risk model.

If a reasonable list of common elements has been compiled, then the next step is to engage the group in attempting to generate a concise (i.e., short) statement that captures the salient common points. This process does not have to result in a single statement. More than one problem may be recognized; however, bear in mind that separate problems may require separate risk assessments. It is the facilitator's job to be sure that all contributing parties in such a process are recognized and that all ideas are given a fair hearing. In this case, we may settle on the statement

What is the cash-flow situation and NPV of a selected plant expansion scenario?

If a high-level and general risk model is constructed, then several, if not all, plant-development scenarios will be able to be evaluated with the same model. This is now the challenge.

SORT IT OUT

If no, or very few, common elements are identified (the case when people have VERY different ideas as to just what is the problem), then a sorting mechanism must be used to arrive at the most important of the issues expressed. This sorting can take may forms. A hierarchy diagram (see selected reading list for this chapter) is a commonly used tool. In a typical hierarchy diagram, the "policies" are listed at the top, the "strategies" are listed in the middle, and the "tactics" at the bottom. The typical form for a hierarchy diagram is a triangle. The sorting mechanism I use is somewhat different. During a break in the meeting, the facilitator should copy all of the sentences and ideas listed on the board to a piece of paper and make a copy of the paper for each participant. When the meeting resumes, each person is asked to label each statement as either a "focus" item or a "not considered now" (NCN) item.

Statements labeled as NCNs can include items that we will accept as fact. For example, someone may list "We must realize returns in excess of our costs." Although this is a good guiding principle, it likely will not be part of the expression of the final question. NCN items can also include details. These are items that may eventually be addressed, but not as part of the primary problem. For example, someone may state "We must train our people in food-processing regulations." This may be true if we decide to implement the food-processing strategy; however, if we do not implement that strategy, training is not necessary. Statements labeled as "focus" items are those that we think are the important problems upon which we should focus our attention. These are the statements that will eventually form the essence of the problem to be solved.

After each person has labeled all statements with one of the two labels, the papers are gathered and those statements that were identified as "focus" statements by more than about half the participants are printed on a new paper and again distributed to the group. If there are relatively few statements that were identified multiple times as focus items, then it is expedient to resort to the common-element process described previously. That is, can we express the elements of the few focus statements as a single problem to which (nearly) everyone can ascribe? If so, then the statement of the problem should be generated from the few focus items. If, however, the number of focus items is great or diverse, then a second round of labeling only the statements previously identified as focus items is undertaken. This time, the people in the group are asked to sharpen their focus and again label each item as either an NCN or a focus item.

In particularly difficult situations, several iterations of this process may be required to arrive at a small set of statements. The small set of statements should either be reduced to a single expression of the problem by identifying common elements in the statements, or should be considered to be separate problems if no common ground can be found. In the latter case, the facilitator should make the group aware that separate risk assessment models may be required for each statement and that the group will have to decide in which order the individual problems will be addressed.

GROUP DYNAMICS AGAIN

Sorting schemes such as that described here are valuable tools, not only for arriving at consensus with regard to just what is the problem, but for fostering good group dynamics. When an individual in a group contributes an idea or statement, that person would like to feel as though his/her contribution has been given a fair hearing and consideration (no matter how inane the contribution may seem). Using these sorting techniques, the contributed ideas that do not seem to apply need not be singled out as inappropriate.

The sorting process described will discard less useful contributions along with many others of greater merit. In this way, the feelings of people who have contributed tangentially useful ideas are not singled out, nor do they feel criticized for their contributions. It is an important part of the facilitator's job to keep as many people contributing to the process as long as possible. Tools like the sorting mechanism help very much in this regard.

UNITS

Following succinct definition of the problem to be solved, the next item to be decided is just how, in the end, the answer will be expessed. This may seem like a technical detail, but believe me, it is not. It will guide the rest of the entire risk assessment exercise.

Typically, if the question is well defined, a discussion of units will follow. For example, if we are considering the problem of adding production capacity to an existing plant, we may wish to express the final answer in tons of product, or as throughput of material, or as a financial measure such as cash flow, or after-tax income, or net present value, or any of a host of other financial measures. In any event, the units of the answer must be decided upon in the beginning of any risk assessment.

In the next section, I will discuss in detail what I call "overarching categories." Examples of such categories may be "Politics," "Environmental," "Technical," and other considerations. Each of these categories typically is broken down into numerous variables. For example, categories and variables might be

- Economic (category)
    Price (dollars-per-gallon) of product
    Advertising costs
    Litigation costs
    Construction costs
    Tax rate
    Operating expenses
- Environmental (category)
    Probability of a spill
    Attitude of local government toward pipeline construction
    Previously existing environmental damage
    Availability of emergency cleanup equipment
- Technical (category)
    Capacity of new pipeline
    Capacity of existing facilities

It is important to establish the units of the "answer" because all variables that contribute to the solution of that answer must, in the end, be able to be combined such that the result is in the desired units. For example, the answer variable may be "after-tax income in dollars." In this case, we must be able to combine in our risk model variables such as "probability of a spill," "attitude of local government toward pipeline construction," and all other variables in such a way as to result in cash flow in dollars.

So, knowing that the end result of the analysis has to be expressed in dollars, we must keep in mind when discussing the "probability of a spill" variable that the probability has to be multiplied by some dollar amount. For the variable "attitude of local government toward pipeline construction," we may decide to integrate this into the risk model by asking two questions such as: "What is the probability that the local government will stop the project altogether?" and "What is the per-week cost of delay if the local government holds up construction?" This variable might also be handled by implementing a unitless scale that can be translated into dollars. For example, we may wish to express the attitude of local government toward such projects on a scale of 1 to 10 with 1 indicating a "good" attitude toward such projects and 10 indicating a "bad" attitude. You may then in your risk model use this 1-to-10 value as a multiplier (or as a pointer to a multiplier) for a value or distribution of values which represent, for example, the weekly cost of local government delays.

It should be clear from the examples listed here that the definition of the output or answer variable and the units in which it will be described are important. This must be known prior to defining categories and variables for the risk model. Certainly it must be taken into consideration when phrasing questions that will be used to prompt information providers.

OVERARCHING CATEGORIES

Every problem is composed of component parts. In the risk business we generally think of these component parts as variables. We do this even though some of the components, in the end, will not vary (constant values or chances of failure, etc.). Thus, I use the term "variable" here in the loosest sense.

The variables that comprise a risk model generally can be categorized. In constructing a risk model, I generally approach the situation by first defining what I call "overarching categories" (OCs). Later I help the group to generate the variables that collectively define each category.

For example, in the case of a plant-construction risk model, I would next challenge the group to define categories for consideration in the risk model. When building a plant, certainly you would have to consider aspects in at least the following categories:

- Political
- Environmental
- Labor (union)
- Technical
- Commercial
- Financial

and probably others. When making the list of categories, be sure to generate a definition for each category. For example, when we list "Labor" as a category, just what do we mean? If we are going to consider just the cost of labor, then why would that not be considered under the Financial category? Considerations such as the availability of qualified labor, negotiations with local unions, and other labor-related considerations may justify Labor as a category separate from the Financial one.

Categories are useful for organizing variables and for orchestrating discussion of how the risk model will be constructed. In addition, categories help to organize presentation of the risk model to the end user. When risk models are constructed, variables typically are added to the user interface in an order that generally is dictated by their appearance in equations or by the order in which they arose in discussion of the model. Categorizing the variables allows for a logical grouping and flow of information in the model user interface.

6 Build a Contributing-Factor Diagram

CONTENTS

The Contributing-Factor Diagram — Getting Started 51
Identify and Define Variables 56
Ask the Right Question 57
Double-Dipping 58
Double-Dipping and Counting the Chickens 59
Fixing the Double-Dipping and Counting Chickens Problem 59
CFD-Building Example 60
Short List of Hints for Building a CFD 65

THE CONTRIBUTING-FACTOR DIAGRAM — GETTING STARTED

As the risk model takes shape, variables that contribute to the solution of the problem will be added. Names for these variables should be agreed upon and the names recorded on some type of movable object such as a large Post-it™ piece of paper or a magnetic "octagon" (the type used in systems-thinking exercises). These name-bearing objects should have on them sufficient room for recording of the name and some additional information. Post-it™ notes are relatively inexpensive and can be "stuck" on a vertical board for viewing and arranging. These arranged items can comprise a contributing-factor diagram (CFD). An example of a CFD is shown in Figure 6.1

The CFD of which I advocate the use is an outgrowth of the more formal and rigorously constructed "influence diagram." In the classical and academically adherent influence diagram, a mix of symbols are used to represent nodes of various type. Decision nodes typically are depicted as rectangles. Ovals indicate uncertainty nodes and octagons signify value nodes. Double ovals or octagons represent deterministic nodes.

Years ago I used to endeavor to construct, strictly speaking, influence diagrams when attempting to outline the system of equations for a risk model. Almost invariably, it turned out that in the initial stages of a project I knew so little about how the various components would, in the end, interact that it was folly to attempt to deduce the esoteric linkages between the myriad puzzle pieces. Even now, after many years of experience in building such diagrams, I find I am doing really well if I can but construct a rudimentary arrangement of the fundamental project components.

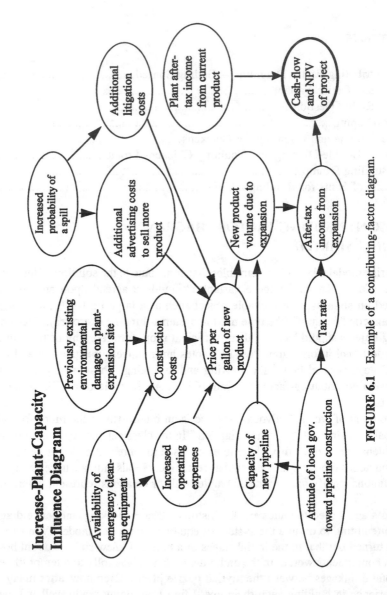

Increase-Plant-Capacity Influence Diagram

FIGURE 6.1 Example of a contributing-factor diagram.

In a CFD, therefore, I make no veiled attempt to suppose that I have insight or foresight enough to deduce the nuances of the risk-model algorithmic relationships. The plain and unadulterated ignorance typical in the initial stages of risk-model building generally precludes the fabrication of a formal influence diagram. This leads to the use of the much simpler, for lack of a better term, contributing-factor diagram.

Before influence-diagram-huggers renounce me as a risk-heretic, I would here like to adduce that influence diagrams certainly do have a prominent role in modeling! Such diagrams can be used for at least two purposes. One common application is to use the plots to outline a work process. Another application is to utilize the influence diagram to delineate the logical and mathematical relationships between risk-model variables. It is for this latter application that I recommend the simpler CFD. A detailed plant-construction example will be given later in this chapter.

Something to realize about a CFD is that it is not a flow chart. In a flow chart, entities are arranged in sequence. Things that happen first generally are arranged at the top of the flow chart, and items that are subsequent are listed below. In contrast, the items arranged on a CFD are not time- or sequence-dependent. Items are connected with arrows that indicate that one item is used in the calculation or determination of another. For example, in Figure 6.1, the variable "Probability of Spill" is used in the calculation of advertising costs and litigation costs. The fact that these variables are posted near the top of the diagram has nothing to do with whether, in the risk model, they are used prior to or after any other variables shown.

In a CFD we generally work backward from the answer. That is, having previously succinctly defined the question and the answer variable and its units, we post the answer variable first on the CFD. Before proceeding with the posting of additional variables, however, there is one more answer-variable attribute to be discussed. The next step in CFD construction is to decide just how the answer to that question will be presented. In Figure 6.1, the answer variable is "After-Tax Income." We will express this in dollars. At the time of CFD construction, the group must come to consensus concerning just how the output variable will be displayed and presented to management. Later in this book, it will be disclosed just what is a distribution. However, it is important at this juncture to consider some practical aspects of distributions.

In Figure 6.2, the variable "After-Tax Income" is displayed by three different bar charts. The data comprising each bar chart are identical, the only difference being the number of bars we selected into which the data would be subdivided for display. It can be seen that when only five bars are selected, the bar chart has a roughly log normal appearance (skewed to the right). When 10 bars are selected, the log normal character is lost; it appears that there is a part of the distribution where the frequencies are all similar and that there are outlying values on either side of this range. When 40 bars are selected, the bar chart takes on a bimodal appearance. These three bar charts of the same data distribution certainly would convey different messages to those not well grounded in statistics.

The three different "pictures," however, are not the worst of it. Consider the situation in which project economics dictate that if the project can't generate about $5 million, then we likely will not pursue it. In the 5-bar chart, the probability of

FIGURE 6.2   Frequency plots of common data using 5, 10, and 40 bins.

generating "around" $5 million is about 50% contrasted with the 10-bar probability of only around 25%. The 40-bar probability is less than 9%. It should be obvious to those versed in statistics that the differences in probabilities are related directly to "bin" sizes. Managers to whom the bar chart of "After-Tax Income" is presented, however, may not be so savvy in statistical inference, and certainly will not be presented with multiple plots of the same data for comparison. This topic is addressed in greater detail in Chapter 10.

At the time of CFD construction, it is essential that the group decide on just how much detail (how large or small the "bins" need to be) needs to be conveyed in the bar chart. Group members should come to consensus on this issue and record exactly why they decided on a level of resolution. All members of the group should agree to consistently present the "After-Tax Income" data.

Following resolution of the problem of how answers will be presented, additional variables should be added to the CFD. These additional variables should be those that contribute to the calculation of the answer variable and which have passed the tests outlined in the Identify and Define Variables, Ask the Right Question, and Double-Dipping sections of this chapter.

For each variable added, the group should agree on a written definition of the variable and just what will be asked of the user. For example, in Figure 6.1, one of the variables listed on the initial contributing-factor diagram is "Attitude of local government toward pipeline construction." At first blush this may seem like the question to ask. Consider, though, that we have defined our output variable to be "After-Tax Income." Therefore, we are going to have to gather from the user information relating to local government attitude that will allow us to include this information in a calculation that results in "After-Tax Income" expressed in dollars.

Upon further consideration, we may decide that "Attitude of local government toward pipeline construction" is not exactly the question we need to pose to the user that will supply data to the risk assessment model. In this case we may want to decompose the original question into the following set of questions:

- What is the probability that local government will delay pipeline construction?
- What is the likely length of time of any delay?
- What is the cost per unit time of the delay?

Users would be prompted to supply the answer to the first question as a distribution of probabilities expressed in percentage. Units for the second question related to time may be a distribution of months. A distribution of dollars-per-month would suffice as input for the third question. The user-supplied data then could be combined thus:

$$(\text{probability of delay}) \times (\text{length of delay in months}) \times (\text{cost per month}) \quad (6.1)$$

In this way, political considerations can be used to contribute to the calculation of "After-Tax Income."

In addition to generally posing each consideration so that it can be used in the calculation of the identified output variable, it is essential that the group agree upon the exact wording of the question/statement that will prompt the user to enter the desired data. For example, for the first question listed above, the group might settle on the phrasing:

> Enter below a distribution of decimal percentages that represents the probability that local government will delay the project due to their attitude toward pipeline construction in the project area. This probability should not include consideration that the government will cancel the project altogether, only that the local authorities will delay the project.

Wording of the question is critical. In this question, it is clear that we want probabilities of delay related only to the government's attitude toward pipeline construction in the project area. This should not include, therefore, consideration of other reasons the company might have for delaying the project (lack of funds, availability of materials, etc.). In addition, we may wish to capture separately the probability that the government will cancel the project altogether. This is a different question relative to calculating the cost of a government-imposed delay. From this simple example it should be clear that the wording and phrasing of questions are critical to gathering the desired information from users.

Pertinent variables are added one at a time to an increasingly detailed and evolving CFD. As each is added, it is placed on the diagram in a position that indicates its relationship to variables already existing on the diagram. Arrows are drawn from the newly added variable to variables that use the newly added variable in their calculation or determination. A poorly designed CFD can appear like a plate of spaghetti with intersecting and crossing arrows. In a well-designed diagram it is easy to trace factors backward from the answer variable to the most important input variables. A well-designed CFD is a blueprint for generating the equations that will form the basis for the risk assessment model.

IDENTIFY AND DEFINE VARIABLES

Following the establishment of categories and the posting of the answer variable on the CFD, the next step is the identification of variables that comprise the categories and the establishment of the relationships between the variables. The risk-process facilitator should guide the group through a discussion of each of the categories to establish just what variables might be appropriate. As outlined previously, we might establish (at least) the following categories and variables for our plant-construction project:

- Economic (category)
    - Price (dollars-per-gallon) of product
    - Advertising costs
    - Litigation costs
    - Construction costs

        Tax rate

        Operating expenses

- Environmental (category)

        Probability of a spill

        Attitude of local government toward pipeline construction

        Previously existing environmental damage

        Availability of emergency cleanup equipment

- Technical (category)

        Capacity of new pipeline

        Capacity of existing facilities

Just as categories can be broken down into sets of variables, the variables themselves can be subdivided. For example, it may be necessary to subdivide the construction cost variable into component variables such as heavy equipment rental costs, compressor (purchase) costs, and so on.

It is the job of the facilitator to ensure that any given subject or variable is not unnecessarily subdivided. Too much detail is detrimental, and the group should strive to keep variables as general and "high level" as possible. It also is essential that any question devised to query a user for variable-related information be posed such that it prompts the user to provide information that will best serve the purpose of the risk assessment.

ASK THE RIGHT QUESTION

Asking the right questions (in addition to correct phrasing) can be critical to success. I have, in the past, been asked to consult with groups charged with supplying clients with risk assessments for construction costs. For one of these jobs, the fundamental problem was that actual project costs deviated significantly from those predicted by the risk assessment model run prior to project inception. Analysis of the risk assessment process employed by this group revealed several problems, not the least of which was the phrasing of questions that prompted users to supply information. Some of the questions posed did not contribute to the solution of the actual problem.

The ultimate goal of the group's risk model was to predict, prior to beginning the project, the total project costs. One of the costs of such a project is that of compressor purchases. Because the risk model was intended to be run prior to the project's actual beginning, the risk model asked, among other things:

- How many compressors will be needed?
- How much does a compressor cost?

The user generally entered a single value (say, 1) for the answer to the first question and entered a distribution of costs (say, $100,000 plus/minus 10%) for the answer to the second question. These costs and calculated total costs were being generated prior to beginning the project. Therefore, projects could still be rejected because the project absolutely was too costly or because it cost more that a competing project. For these reasons, there was considerable pressure by project proponents to minimize costs. The costs for compressors are fixed by manufacturers, so there is little room for cost savings there. However, it could be proposed that a small number of compressors will be needed, thus keeping costs down.

The actual number of compressors needed for a given type of project generally increased through the life of the construction project relative to the number entered in the pre-project risk assessment model. Therefore, it should come as no surprise that the actual total cost of a project significantly exceeded initial pre-project estimates. To remedy this problem (in part), I suggested that they ask different questions with respect to compressor costs. The questions I proposed they ask are:

- When projects of this type are finished, how many compressors did we typically use?
- How much does a compressor cost?

Asking these questions will yield much more accurate total-project costs, with respect to compressors, than the original set of questions. Much of a facilitator's job centers around limiting the amount of detail (yet capturing sufficient detail) demanded by the model. In addition, the facilitator should be concerned with asking the right questions. Pertinent questions should be phrased such that answers to questions result in user-supplied information that actually contributes to answering the question that is the aim of the risk assessment model.

DOUBLE-DIPPING

Generally, the facilitator is much less familiar with the problem than are the members of the group. However, a facilitator should use common sense and his/her relative ignorance to advantage. Realize that no one else in the room can ask "fundamental" questions with impunity. The facilitator should question the addition of each variable and consider whether or not the information that is to be captured by the new variable is not, in fact, already captured by some other variable or set of variables. Avoidance of "double-dipping" (capturing essentially the same information more than once, but perhaps in different ways) always is a major concern in model building. Often, group participants are "too close to the problem" to realize some of the overlaps in logic. This especially is true when representatives from different disciplines approach a common problem.

For example, it may be deemed that operating expenses are too high. An engineer may attempt to reduce operating expenses by suggesting that certain types of more expensive but more durable materials be used in the operation, thus lowering operating expenses in the long term. The economist, however, has already in his/her mind discounted the use of those materials because they are too costly. The economist's approach to the problem is to add a third shift, thus reducing overtime expenses. Both approaches solve a common problem (operating expenses a bit too high). However, we would like to keep construction costs down (use cheaper materials if possible); and we would like to keep the number of employees to a minimum (2 shifts better than 3). Thus, we likely would invoke one of these operating cash-reduction methods, but not both. It generally falls to the risk-process facilitator to ferret out the duplications. Getting each participant to express *why* he/she is suggesting a particular approach for solution to a problem and employing experience and common sense are typical ways to guard against double-dipping.

DOUBLE-DIPPING AND COUNTING THE CHICKENS

Double-dipping appears in many insidious forms. One particularly menacing embodiment of the malady is what I like to call the "counting the chickens" problem. This topic is covered in more detail in the "Type" Consensus Models section of Chapter 3, but because of its link to double-dipping, it bears mentioning here.

Double-dipping is a difficult hazard to avoid when building a model. A good facilitator attempts to ensure that no piece of information is collected, in different ways, more than once in a risk model. Avoidance of this problem is, more or less, controlled by the facilitator and the model builders. However, once the model is built and "shoved out the door" to the user community, double-dipping of a different type can take place.

Only organizational and process-related safeguards can even begin to avoid the "counting the chickens" problem. Consider that we have five houses that form a circle. They, therefore, share a circular back yard. In the yard there are a dozen chickens. We knock on the front door of each house and ask the occupants how many chickens they have. They look out the back window and report that they have 12. This same answer is given at every house, leading us to believe that 60 chickens exist. Such counting of things more than once is a common failing in the organizational implementation of a risk assessment process.

FIXING THE DOUBLE-DIPPING AND COUNTING CHICKENS PROBLEM

The counting-the-chickens problem is an especially difficult dilemma to solve. There exists, to the best of my knowledge, no technical basis for its solution. Organizational awareness and aplomb tend to be the only remedies. Although a potential solution, organizational cures are not easily implemented.

Consider a company that produces a product in several locations around the world. At each location the company expects that the lion's share of the product will be consumed by the local indigenous population. The volume of product produced in excess of the amount consumed locally is expected to be exported to a targeted population of consumers. It is not uncommon to find that multiple product-production facilities have based their business plans on a common population of consumers. Those consumers might be able to absorb the excess production from one facility, but certainly not from all plants. Well, in this situation, how do we prevent the counting-the-chickens (i.e., counting the same consumers more than once) problem?

As previously stated, there likely is no simple or even intricate technical solution to this problem. That is, implementing a software or hardware solution probably is without merit. Organizational processes probably are the only avenues to a possible remedy, but organizational solutions can be notoriously untenable.

For example, the aforementioned product-producing corporation might decide that it will now require that all such proposals be funneled through corporate headquarters prior to actually implementing the plan. At corporate headquarters, then, it must be somebody's job to investigate each proposed production plan and to ferret out the overlaps in plans such as counting potential customers more than once.

This solution works well when the data containing the proposed plans, sent from the producing locations, are "truthful" and accurate. Exporting is more expensive than local consumption, so the plans may "play down" the amount of exports. This can lead the corporate overseer to believe that the consumers targeted to absorb exports are capable of "taking up the slack" for more production facilities than they actually can. In addition, the corporate overseer position often is not a popular one. It tends not to be a career-path position and is a post that is viewed with some degree of resentment by the operating facilities that are being "governed." Therefore, this position typically experiences high turnover; that is, those assigned to the job generally take the first opportunity to vacate the position. Without consistency with regard to staffing of this critical assignment, much of the effectiveness of sleuthing out problem areas is lost.

Another popular remedy is to assign a person or group to temporarily become part of the production-site team that is going to put forward a proposal. The assigned person generally "moves" to the production site and becomes part of the process and organization that will generate the proposal. The idea here is that such a person who joins multiple teams in different locations will be able to spot the problems before they become part of the proposal. This approach, however, has some obvious drawbacks.

First, the employees whose job it is to join the location-specific team must be willing to move to the remote site for an extended period of time (and repeat the process when the current assignment has ended). This can definitely be done, but it can't be done indefinitely. In addition, more than one proposal might be fabricated at the same or different facilities. This requires that there be more than one employee assigned to such a job. If more than one person is involved, then avoiding the counting-the-chickens problem depends upon close coordination and good communication between those individuals. This, in itself, can be difficult to implement.

Although both solutions outlined here are only partial cures and are far from perfect, implementing some manner of organizational control to avoid "double-dipping" is better than instituting no plan at all. The counting-the-chickens problem is a particularly insidious risk-modeling problem and can wreak havoc on a company if it is not suitably addressed.

CFD-BUILDING EXAMPLE

Let's consider a situation in which Joe has to build a new chemical plant. In the initial planning stages, Joe might wish to construct an influence diagram that will depict the stages and steps that need to be taken between the present time and the event of opening of the plant. In such an influence diagram, Joe would begin by posting the "answer" variable — the opening of the plant. Joe knows that at least the following variables will influence the plant opening, and, therefore, need to be included in his diagram.

- Negotiations and contract signing with the labor union (negotiations may not be successful).
- Identifying local suppliers for building materials. There is no guarantee that critical supplies can be had locally.
- Governmental approvals and permits must be obtained. Permits have, in the past, been denied for similar projects.
- Engineers must submit a construction plan that not only is economical, but that will comply with environmental regulations. In addition, the plan must result in a plant that will fit on the site but will provide sufficient production capacity. This is not a given.
- Local environmental groups must be placated.

Joe will commence with his project by organizing the aforementioned items and others as an influence diagram (not shown here). Rectangles (decision nodes), ovals (uncertainty nodes), and octagons (value nodes) will be used to represent the individual items. For this type of planning, influence diagrams are invaluable.

Ultimately though, Joe will be faced with the arduous task of calculating, for example, the net present value (NPV) and perhaps the internal rate of return (IRR) of the plant. In the initial stages of the project, Joe does not have a firm grip on exactly what variables will contribute precisely to a set of equations. He does have some inkling as to the general relationships between the variables that will be used to construct the strict set of equations of the risk model. For this purpose, at this stage, a contributing factor diagram (CFD) is more suitable.

In a typical CFD, all variables are represented by a single common symbol — an oval, a rectangle, or whatever. No intrinsic meaning is attached to the symbol. Symbols are, like those in an influence diagram, connected by arcs and arrows to indicate which variables contribute to the calculation of other components (even though, at this stage, the precise equations that might be used to calculate a given variable might not yet have been decided). An initial stage of a CFD is shown in Figure 6.3.

In Figure 6.3, Joe has posted the "answer" variable (NPV and IRR of the new plant) and several major factors that will contribute to the calculation of NPV and IRR. Joe knows that he will have to calculate Total Cost and that, at least in the initial stages of the project, Construction Cost will be a major part of the Total Cost. He knows also that to have "healthy" NPV and IRR values, he must more than offset the costs by generating sufficient income from the plant.

In a meeting with a facilitator and all other pertinent players in the plant-construction project, Joe presents his initial CFD and asks for comments/additions/etc. One of the first things to be pointed out is that the Construction Cost variable needs to be subdivided into its major component parts. The facilitator, in

FIGURE 6.3   Initial contributing-factor diagram.

an attempt to keep things simple, challenges the breakdown of the Construction Cost parameter but is quickly convinced that at least some subdivision is pertinent.

Subdivision proponents argue that if the plant project is killed in the initial stages by abnormally high construction costs, it will not be sufficient to know, and to send to a database, simply the fact that construction costs were exorbitant. It will be necessary to know whether labor or steel or concrete or whatever parameter (or combination of parameters) causes the Construction Cost variable to contain coefficients of unreasonable magnitude. As can be seen in Figure 6.4, the original Construction Cost variable will represent the combination of costs related to labor, pipe, steel, compressors, concrete, and exchangers.

Labor, it is decided, will be considered as two types. Construction labor is considered a capital expense (CAPEX) while "operating" labor (plant workers, office staff, etc.) is considered an operating expense (OPEX). Later in this book in the section Comprehensive Risk Assessment Example in Chapter 14, I will use this CFD to develop the associated risk model. It will be realized at risk-model-building time that another variable that apportions CAPEX and OPEX labor costs on a year-by-year basis will need to be added as a risk-model parameter. At the current stage of development, however, Joe and his cohorts are unable to predict all the variables and relationships that will be required by the equations of the risk model — thus the CFD.

FIGURE 6.4 Completed contributing-factor diagram.

Labor, it is decided, will be calculated from the factors Days per Year, Hours per Day, Hourly Wage, and Number of Workers. This will be done regardless of whether Joe is calculating CAPEX labor costs (arrow between Labor Costs and Construction Costs — Yearly) or calculating OPEX labor costs (arrow between Labor Costs and Total Cost per Year).

Joe's group decides that Days per Year not only contributes to the estimation of Labor Costs, but also will serve as a term in the equation that calculates Daily Plant Income. The Dollars per Ton and Tons per Day variables also will serve as terms in the equation that calculates Daily Plant Income.

Variables Daily Plant Income and Corporate Tax Rate will be the major contributing factors in the calculation of Total Income per Year. Along with Discount Rate, this answer parameter, represented as a time-series variable, will, in turn, be the major contributing factor in calculating yearly cash flow for the NPV and IRR estimates.

In addition to Labor Cost OPEX and Construction Costs, Joe's team decides that the other major contributing factor in the calculation of Total Costs is Utility Costs per Year. At the time of the initial CFD-building meeting, Joe's team decides that two factors should contribute to the calculation of Utility Costs/Year. Kilowatt Hours per Year and Cents per Kilowatt Hour are added for this purpose. However, it will be discovered later that the team decides that if utility costs are exorbitant, there is no real information, all things considered, in knowing whether the number of hours or the cents per hour is the culprit. In addition, the team, through use of the model, will discover that users are more adept at supplying total energy costs per year (as you might be if you were queried concerning your home's energy costs) and are less adept at providing accurate and precise information about hours and cents.

As you will also discover in the Comprehensive Risk Assessment Example of Chapter 14, the team will decide that the Construction Costs should be treated in a separate model. The output distribution from the Construction Costs model will be used as an input distribution for the main NPV/IRR-calculating model. The team decides this because the construction phase of the plant is imminent and the construction teams, at this stage, have to rerun the model multiple times in an attempt to emulate several construction scenarios. In addition, this cadre of construction consultants has a pretty good grip on the associated costs, etc., but they do not have a clue concerning items such as Discount Rate, OPEX costs for labor, Dollars per Ton, and other variables. They would be obligated to supply values for such variables if the model is not divided.

One more thing to note about the CFD in Figure 6.4 is that each variable, starting with the answer variable (NPV/IRR), is labeled with units. A good facilitator will preclude the addition to the CFD (not an influence diagram) any variable that can't be put into a set of equations and somehow be used to calculate the answer variable. This is an important point because the risk-model builder likely will not have all team members available at the time of building the model and will have to rely on the CFD as a "blueprint" for model assemblage.

SHORT LIST OF HINTS FOR BUILDING A CFD

The numbered steps below are a synopsis of how to approach the building of a CFD.

1. Decide what the "answer" variable is to be (this can be difficult!). There does not have to be just one "answer" variable, but decide on just one to start with.
2. Decide in what units you want the answer variable to be expressed. For the sake of this example, let's consider that you want to calculate NPV and that you want it expressed in dollars.
3. Begin to add "large" variables to the diagram that would contribute to NPV. By "large" I mean things like "costs," "incomes," etc. Then, for each "large" variable, show the things that go into making up the "large" variable. For example, "costs" might consist of two variables called "OPEX" and "CAPEX." OPEX might consist of labor costs, rental fees, etc. CAPEX might be composed of the cost of steel, the cost of land, etc. Likewise, "incomes" might include "number of barrels sold," "price per barrel," etc.
4. For each variable that you add, be **sure** to decide in what units it would be expressed. For example, if you add a variable to the diagram called "Politics" because you think there might be political problems with the project, consider that your output variable (NPV) is expressed in dollars. How would you include "Politics" in an equation that would result in dollars? Well, I would change the "Politics" variable to three new variables. They would be
   - The probability (percent chance) that we will have a political problem (expressed as a distribution of values between 0 and 1).
   - The length of time we might be delayed if we have a political problem (expressed as a distribution of days).
   - The dollars per day it will cost us (expressed as a distribution of dollars) if delayed.

   Now I can multiply the percent probability of delay times the length of delay times the dollars per day and get a dollar value that can be used in an equation to calculate NPV.

   Likewise, don't put a variable on the contributing-factor diagram that says, for example, "Electricity." What you really want is "kilowatt hours" expressed as a distribution of hours and "cents per kilowatt hour" expressed as a distribution of cents. These can be combined to influence NPV (in dollars).

If you use the logic outlined above, you should end up with a contributing-factor diagram on which each variable can contribute to an equation or set of equations that will, eventually, calculate cash flows and then NPV.

7 Education — Critical at Three Levels

CONTENTS

Education for Management 67
Introductory Class for Risk-Model Builders.. 69
Advanced Training for Risk-Model Builders.. 70
Organizational Attitude and Training Centers 71
Time Demands and Organization of the Education Effort.................................... 72

EDUCATION FOR MANAGEMENT

Without exception, the most critical element in risk-process implementation is that of education. As will be delineated in the following paragraphs, education at (at least) three levels is necessary. The three levels are presented separately in three sections of this chapter.

There is not much point in training the "troops" to think stochastically and to generate sophisticated risk models if the decision makers (usually management) are ignorant with respect to the nuances of and output from the risk process. A risk-savvy employee is not doing anyone any favors by plopping down on the unsuspecting manager's desk a menagerie of probabilistic plots, risk-weighted values, and other not-so-easy-to-understand stuff. In any organization, if the decision makers (1) are not prepared to deal with the input/output from a risk assessment, (2) do not know what questions to ask, and (3) do not know what to look for and what to look out for, etc., risk-wise employees can expect embarrassed and feeling-put-upon management to coldly receive any foreign-looking risk-related information.

Employees should not expect unprepared managers to warmly embrace their stochastic overtures. In addition, the purveyors of the risk process should not expect managers to sit still for multi-day classes and workshops in which are conveyed the many details of risk-model building. Education of managers with respect to risk assessment is a special process that must be handled deftly and with some organizational and psychological aplomb.

I have learned the hard way that any course designed for managers must be short in duration — usually no more than 1 day. That is not to say that special department-specific seminars in "retreat-type" situations should be discouraged or discounted. However, if a large number of managers is to be trained in the ways of risk, a course must be designed that can be inserted into a busy schedule with minimal interruption. Usually 1 day can be set aside. More than 1 day becomes a problem.

It also is important that a risk class for managers be able to address any number of attendees. This is practical because there are occasions when a very small group of managers (sometimes just one) from a specific organization will request such training. Therefore, any class that involves "role playing" or interaction between class attendees likely will be less applicable. Any exercises offered in the class should be able to be performed individually. Attendees should not be required to have out-of-the-ordinary computer or other skills.

It should be pointed out at the onset of the class that the risk-for-managers class is not a "dumbed-down" version of a "regular" risk class. This should not only be said, but it should be true. In my classes for managers I always present the analogy of airplanes or ships and the people associated with them. There are those who build planes and ships. Those people need to understand all of the details about how an aircraft or ship is constructed. However, it is unlikely that you would want any of the builders at the controls on a stormy night. These plane and ship builders are equivalent to those employees who attend the "regular" risk classes in which are conveyed the concepts and practices of risk-model building.

On the other hand, I am not sure you would like the captain of the plane to attach the wings. Her/his role is different. True, such a person has to have sufficient knowledge of construction so they don't make unrealistic demands of the craft, but their primary purpose is to guide the plane or ship safely to its destination. So it is with managers and risk assessment. Most managers will not be able to write the computer code necessary to construct a risk model. However, they should be comfortable with risk-model output, they should know what questions to ask, they should share a common jargon with the risk-model presenters, and they should be capable of making decisions based on the output from multiple risk models. Therefore, it needs to be stressed that the risk class for managers is not a "dumbed-down" version, but rather is a class with a completely different slant than the classes offered to those who will learn to build risk models.

Nothing sells like success. A class for managers should invariably commence, after introductory material, with a parade of examples from across the organizational spectrum illustrating where and how risk technology has been applied and the benefit it has been to the organization. At least two major benefits result from such a "show and tell." First, class attendees will realize that people not unlike themselves have put this process to use to great advantage. Reduction of apprehension is key. Second, some of the class participants will have come to the class wanting to know about risk assessment, but really have no idea how they might use it in their organization. A broad-based demonstration of what others have accomplished should be designed to allow them to "see themselves" as potential participants in the risk initiative.

Because a class for managers likely will be given often and because the potential audiences will represent quite diverse entities, it is impractical to "tune" more than just a small part of the class to directly address the specific problems of any given group. Therefore, examples shown and exercises performed by the class members should be as generic as possible. Lecture and exercises should be geared toward imparting the fundamentals of the risk process, such as

- What is risk?
- How does it relate to decision making?
- What is uncertainty?

- What is a contributing-factor diagram?
- What is a distribution?
- What is Monte Carlo analysis?
- What is a frequency plot?
- What is a cumulative-frequency plot?
- What is chance-of-failure and why is it important?
- What is dependence and why is it important?
- What are risk-weighted values and what can I do with them?
- What is sensitivity analysis and why is it important?

It is important in such classes to avoid getting sidetracked into discussions of *how* most of these risk-related items work. For example, a discussion of how a distribution is built or how a Spearman rank correlation is calculated should not take up class time. Emphasis should be put on what these things are, why they are important, and what managers can do with them.

At the conclusion of the class, managers should be given a short list of items that they can pass along to those who might present risk assessments to them. It is suggested that this list be given to all potential presenters well in advance of any risk assessment presentation and that the presenters be prepared to discuss each of the items on the list. This is a powerful list and process. Even if the manager has by the time of the presentation forgotten some of the risk-related details, presenters will not show up unprepared. They will have "done their homework" with respect to generating a credible risk analysis. The list also aids in consistency. With the list, everyone knows just what is expected, and similar information is presented by diverse groups for a wide range of projects. This greatly aids the manager in the process of comparing and ranking projects of similar or different types. A list might look like that shown below.

When reviewing risk assessments, demand

- To have described just what is the question being answered by the risk assessment — demand specificity;
- Consistency in evaluation and a consensus model if multiple opportunities are being compared;
- To see a contributing-factor diagram;
- A discussion of chance-of-failure and how it was handled;
- A discussion of dependencies between variables (in the contributing-factor diagram);
- A discussion of risk-weighted numbers (if appropriate).

INTRODUCTORY CLASS FOR RISK-MODEL BUILDERS

A second level of education and, thus, a second set of classes has to be designed for those who want to know how to build a risk model, but presently know nothing about it. Like the class for managers, these classes should use examples that are as generic as possible so that engineers, economists, scientists, and people representing other disciplines can relate to the lecture and exercises. Unlike the manager's class, this class should be up to 2 days in length.

Classes aimed at the novice risk modeler need to begin with the very basics. However, unlike the manager's class, care should be taken to delineate, to a practical degree and depth, just how things actually work. For example, just how does the software package you are using generate each type of distribution? Exactly how do you implement dependence, and what equations are used to do so? What are the equations behind the calculation of the various types of risk-weighted values? Through example, lecture, and hands-on exercises, class attendees should gain reasonable understanding and practical experience related to all fundamental aspects of risk assessment.

Individual lectures and exercises should be designed to present each salient aspect of risk assessment beginning with the decision-making/contributing-factor-diagram process. Terminal exercises should guide participants through the construction of meaningful financial measures such as net present value. A suggested plan of attack for a class might be as follows:

- Discussion of decision-analysis/making tools such as contributing-factor diagrams, sorting process, and other applications;
- Definition of risk and uncertainty;
- How risk technology is being used in this business and in other businesses;
- Distributions — all about them;
- Monte Carlo analysis;
- Frequency and cumulative-frequency plots;
- Chance of failure;
- Dependence;
- Time-series analysis;
- Sensitivity analysis;
- Risk-weighted values.

Upon completion of the course, attendees may not have the skills to build sophisticated risk models alone — they likely will still need help with that. However, class attendance should produce a crop of users who are no longer intimidated by the concept of risk assessment, who understand the fundamentals of risk technology, who feel comfortable with whatever software is being utilized, and who can construct basic risk assessment models. If your class accomplishes this much, "ya done good!"

ADVANCED TRAINING FOR RISK-MODEL BUILDERS

A third level of training has to be aimed at the relatively risk-astute people in the organization. Those individuals who have taken the introductory course and those who have some previous background in statistics or risk assessment require a special class. For those who have taken the introductory class, this advanced training should reinforce concepts and challenge participants to put to work what they have learned. For those who have not been previously indoctrinated in a company class, this course should contain enough review material to introduce these people to "how we do things," but not so much review as to bore to tears those folks who have attended the introductory class.

The format for the advanced class should minimize lecture and maximize exercises and problem solving. A format I have found to be successful is to provide the class members with increasingly-difficult problems to solve. After each person has individually solved the problem, the instructor should review how she/he solved it and then solicit alternate solutions from the class. Lively and instructive discussions often ensue.

Initial problems might focus on simply having class participants interpret what types of distributions might be used to represent a host of variables in the presented problem. Subsequent problems should introduce complexities such as dependence, chance-of-failure, and other risk-related techniques. Final problems should include time-series analysis and the calculation of financial measures such as net present value, internal rate of return, and others.

ORGANIZATIONAL ATTITUDE AND TRAINING CENTERS

A complete understanding of the training facilities and policies within the organization also is important. This aspect of risk-technology transfer often is not considered in the design of a new process. No new process should be generated without a complete understanding on the part of the technology generator as to how people will be trained to use the product. It is a mistake to assume that potential users will want to use the product or, even if they want to, that they will be able to successfully use the new technology. Training is absolutely vital.

Consideration of the training aspect should come early in the product design. It is not enough that the new process be easy to use. A fundamental error committed by many technology generators is they assume that because the new process is easy to execute, users somehow will have the same fundamental understanding of all aspects of the process as is the risk-technology creator. Nothing could be farther from the truth. The only safe way to field a new process is to make sure that the users fully understand all aspects of the new risk technology. This can only be done through training classes. Never assume that users will read documentation — TELL them and SHOW them.

Introduction of risk technology often requires the concomitant introduction of new hardware or software. It is worth mentioning here the relationship between requiring new equipment for a product and a training program. Training within organizations is accomplished by a wide variety of processes and mechanisms. To implement a successful training program, you must have a good understanding of "how things work" with respect to training facilities and policies. No matter what the process, it is likely that until the product has proven itself in financial terms, the person introducing the risk technology and associated management should expect to have to supply the required new equipment for the training program. If training takes place at the risk-product-creation site, then this likely is not a major problem (although it is more expensive to send multiple trainees to a single site than it is to send a single trainer to multiple sites). Likewise, if receiving-site management has agreed to purchase the required equipment, then training at that site also should not be a problem.

Many organizations handle training through central "training centers." Training centers can be financially autonomous, and it may be difficult to convince them that major funds should be spent to provide training for a new and unproven process. In the case of the resistant training center, there are two practical solutions to this problem. Either the risk-product-creation site has to provide the required equipment, or the product-receiving site has to be convinced that it is in its best interest to fund the purchase for the training center. Receiving sites may buy unique equipment for the new risk technology because they are convinced of the value of the new process, but it will be a rare instance when a training group or facility will make the same accommodation. These practical problems should be considered early in the risk-technology-transfer process.

Some organizations have no training policies at all. If the organization has an established training procedure, it is critical that the risk-process creator insist upon doing the training or (if he/she is, for example, a terrible speaker) that the process creator control the course content and who teaches the class. It also is most practical if the process creator can design the training classes so that they can be taught in available facilities and on available hardware. Training should not be the last thing to be contemplated. When considered at the onset of risk-technology development, training procedures can be woven into the design of the process. Training is essential to the process of risk-technology transfer.

TIME DEMANDS AND ORGANIZATION
OF THE EDUCATION EFFORT

In life in general, the time it takes you to complete your tasks can depend more on your degree of organization than on the absolute amount of work to be done. So it is with the education and training effort associated with launching a risk assessment effort. One of the things that shocks risk-process-proponents most is the amount of time that actually needs to be devoted to the educational aspect.

Astute organization of the educational facet can help reduce the time investment. However, even the most fastidiously organized individuals will attest to the fact that the educational process is "eating their lunch." Shortcuts in the organizational effort can lessen the impact of risk education.

For example, in my line of work I have the pleasure to interact with individuals and groups that represent a broad spectrum of disciplines. Within a short time span, it is not unusual to meet with and teach classes to lawyers, accountants, managers, economists, manufacturing personnel, engineers, scientists, and others. If it becomes your charge to interact with such a large number of diverse disciplines, there are basically two ways you can go with regard to generating training materials or examples.

The first method is to prepare classes that contain detailed risk-related exercises that address specific areas. Using this approach, you present general generic material in each class but design "hands-on" exercises that are discipline-specific.

One advantage to this approach is, of course, that the attendees can relate directly to the material and, therefore, absorb more of what is taught. In addition, class members walk away with practical "go by" examples that they can put to use as soon as they return to their jobs. This approach tends to increase the popularity and demand for the class within each discipline due to favorable word-of-mouth reviews.

It should be clear that the disadvantages of this method are mainly yours. Attempting to generate meaningful (and not goofy) complex, risk-related examples for a multitude of disciplines (or even one if you know nothing about it) can be time consuming and discouraging. This approach also puts the onus on the class developer to keep up with changes and advances in each discipline. Things change rapidly, and it does not take long to have a hard-earned set of examples become passé. Reworking a class can be more time consuming than setting up the class in the first place.

I have known others who have taken the area-specific-example route to hire experts in the various areas. Experts are brought on board to design or help design meaningful class examples. This approach relieves the class presenter of the burden of creating exercises of substance, but usually leaves them with the task of presenting the material. Not many other things you can do will more rapidly deflate your credibility with the class than presenting material with which you are obviously uncomfortable. For example, a room full of accountants will perceive in one second that you really don't understand the material that you are presenting (material that was prepared for you by another accountant). You are the teacher. If you don't understand a subject better than those to whom you are presenting, and if your knowledge of a subject is a "mile wide and an inch deep," it is far better to avoid that situation entirely.

Classes with discipline-specific exercises also can be co-taught by the risk expert and by an expert in a given field. This approach works well when the expert in the field also has a grip on the fundamentals of risk technology. If the expert does not, then their presentation, in the case of the accountants, will tend to be simply an accounting exercise plopped into the middle of a risk class. Experts also should be presenters of quality and be available when you need them. Good luck!

I have chosen to take the course of using the same set, with minor exceptions, of generic and general examples for every class. In each class, I incorporate a mixture of general business, political, environmental, etc., exercises with which I am comfortable and about which I am knowledgeable.

The advantages of this approach are many. First, this practice limits preparation time — each class of a given type draws on the same examples no matter the makeup of the audience. Nobody in the room knows more about the examples than do I (or at least I like to cling to that belief). Because I composed and understand the examples, they can, with a minimum of effort and without having to seek counsel, be updated and enhanced. When presenting material with which I am familiar and comfortable, I can concentrate on a relaxed, humorous, and interactive delivery. I neither fear questions nor am I skittish about delving into the details of a situation.

Another real and practical advantage to this teaching method is that when "on the road" traveling from one teaching engagement to another, using a common set of examples greatly reduces the volume of physical "stuff" I have to haul with me. You won't be able to appreciate this point until you are the subject of those disapproving looks from airline employees who are eyeing your carry-on dunnage.

Using common sets of examples also has its drawbacks. If, for example, the audience is composed of a room full of accountants and none of the class examples deal specifically with accounting, taking this approach counts on each class member's ability to translate the class work to his/her own situation. I attempt to mitigate this foible by asking of the class, after each example, if there is anyone in the audience who can see a practical and immediate application of the just-presented principles to what they do. Such a query never fails to produce a crop of raised hands. Turning the floor over to several class members to relate their viewpoints almost always makes clear the link between the exercise and the discipline represented by the class. An advantage to this approach is that I don't have to have a clue concerning what they are talking about!

Another related drawback is that I am sure that class attendees do not absorb as much of the doled-out information as they would if the class patter were couched in their jargon. Some things you just have to learn to live with.

With regard to examples and exercises, the generic vs. class-specific approach represents one proverbial fork-in-the-road dilemma in which you have to choose which way to go. Another such bifurcation is the choice between regularly scheduled "mass training" sessions and a "call me and I will come to you" approach. Certainly in every training effort there is a mix of both methods, but one must dominate.

Using the mass-training doctrine, the provider of classes generally arranges with training centers in various locations to hold classes on specific days. This schedule, then, is "announced to the world," and those interested in the subject are encouraged to attend.

This approach to teaching has several salient benefits. First, people can see a schedule of classes that extends relatively far into the future (usually a year or more). This allows them to plan ahead and work the classes into their schedules. This also is true for the class instructor. Regularly scheduled classes also help prevent "running around like a nut" on the part of the teacher, trying to bring classes to organizational cul-de-sacs populated by small numbers of bodies.

Picking up on the immediately preceding point, the mass training approach is more efficient with respect to being able to reach the greatest number of people with the least effort. It is not, however, most efficient in other respects that I will elucidate in later paragraphs of this section. A typical "call me and I will come to you" class will have in it no more than 5 to 20 persons. Properly promoted mass-training classes usually have a student count of 100 or more. The presenter's travel is reduced and the impact per trip is greatly enhanced.

Still another advantage of the mass-training method is consistency of the message. If the aim of a training effort is to primarily bring a common jargon and understanding of the subject to an entire organization, then presenting the same material a minimum number of times aids in achieving this goal. A message repeated a great number of times to relatively small numbers of people tends to "wander," despite the best efforts of the presenter to be consistent.

The mass-training approach, however, also has its minus column. With respect to total cost, it likely is not most efficient. Although under the "call me and I will come to you" scheme the presenter makes many more trips, those relatively paltry costs are greatly exceeded by the expenses related to having 100 or more people travel to a common site for a multi-day class. In addition, classes taught at the customer's location can usually be held in a room or facility that costs very little to occupy for a day or two (typically a conference room). The mass-training approach invariably requires bearing the expense of space rental (not to mention computers, projectors, etc.).

Mass training also robs the presenter of the opportunity to experience how various parts of the organization operate. Personal one-on-one interaction is sacrificed.

Another drawback is promotion. In a previous paragraph I alluded to a "properly promoted" mass-training class. Advertisement of such classes can be exceedingly difficult to facilitate. Even with e-mail lists and modern communications, making everyone in an organization aware of the class content and class schedule is a very tough thing to do.

The advantages of the "call me and I will come to you" approach are fundamentally the opposite of the disadvantages associated with mass training. This method is cheaper to enact organization-wide. It promotes personal contact and gives the class presenter the opportunity to visit and learn about the various arms of the organization. On the other hand, this approach to training is hard on the presenter with respect to travel. Over the same amount of time, this teaching method will reach far fewer people than will the mass-training approach.

Through the years I have developed a teaching philosophy that incorporates both approaches, with the mass-training method dominant. I attempt each year to schedule as many mass-training sessions as is practical. When someone or some group calls to arrange a class, I make every effort to divert them to the mass-training class nearest in time and space. If, for whatever reason, they claim attendance is not possible, I then weigh other schedule demands against their needs and make a call. Sometimes I say no. Everyone in the organizational-education game will have to define for themselves some satisfying scheme.

A last point on education. Try not to be amazed at the time and effort required. Generating a lively and cogent class from a blank piece of paper is hard enough (you can say a lot of stuff in an hour). However, the biggest drain is the travel and time it takes to carry out an educational assault. Beware when you start this thing that you are creating a monster. Plan accordingly.

8 Risk Assessment Technologies

CONTENTS

Different Approaches 77
Discriminant-Function Analysis 77
Bayesian Analysis 79
Decision Trees 82
Factor Analysis 87
Neural Nets 89

DIFFERENT APPROACHES

Risk assessment models are of a variety of types. Fate-transport, political, technical, financial, and other varieties of models can be constructed. Regardless of the application, risk can be evaluated by applying a wide range of technologies. Risk assessment itself can loosely be defined as the application of any technique that deals with probability in some technically sound manner. Some of the more popular risk assessment techniques applied are Bayesian analysis, discriminant-function analysis, factor analysis, neural nets, Monte Carlo analysis, and others. Although this book will concentrate primarily on Monte Carlo applications, I will here say a few words about some other popular technologies.

DISCRIMINANT-FUNCTION ANALYSIS

Discriminant-function analysis (DFA) has many uses, but the primary application generally is to define "boundaries" (equations) that separate predefined populations of objects in n-dimensional space. DFA also defines (quantifies) the degree (probability) to which the objects "belong" to the populations. The defined discriminant function(s) can be used to then determine for a "new" object the population to which it belongs and the associated probability (degree to which it belongs to the population). Because DFA has no ability to separate objects into groups, it requires objects (data points) that already are assigned to groups and requires measurements on variables associated with each of the objects. An example might be a group of animals that have been subdivided into nocturnal and nonnocturnal types. On each of these animals, we may have measured height, weight, sex, diet, skin covering, etc. (variables).

Although understood since the publication of R. A. Fisher's work in the thirties, DFA gained popularity as a probabilistic technique only after the advent of high-speed computers. As stated above, DFA typically is used to define the "boundaries" between populations in n-dimensional space and, using one of several distance measures, determine the "degree" to which each member of a given population belongs to that population.

To illustrate the space-partitioning functionality of DFA, we will use a conceptual model. Consider two sacks of potatoes hanging side-by-side in a doorframe. We will designate the sacks as sack A and sack B, each containing 50 potatoes. Each potato represents a data point. If we project the position of each potato onto either the right or left parts of the frame, because the sacks are hanging side-by-side, the projections of the two potato populations completely overlap. Even if we lower or raise one of the sacks half a sack length, the projections of the potato positions on the vertical part of the doorframe still will overlap. In the area of overlap, we cannot be sure if the projection of a particular point is a projection from sack A or B. If, however, we project the potato positions either on the floor or on the top of the doorframe, the A and B populations can clearly be distinguished.

It should be evident that if our sacks took on complex shapes or were oriented in the doorframe in a configuration other than side-by-side, we would have to find another plane in n space (other than the floor or the top of the door frame) that, when projected upon, would clearly separate the two populations. This is the task of discriminant-function analysis. DFA attempts to find a function in n space that, when used as a "backdrop" for data-point projection, will do the best job possible of maximizing the distinction between the populations. As you might imagine, as data-set shapes become complex and as the data sets between which the discriminant function must distinguish become greater in number, the complexity of the discriminant function increases dramatically.

One of the primary objectives of DFA is to "examine" the distribution of data in n-dimensional space (in this simple conceptual example, 2-D space) and establish a "new axis" upon which the data points can be projected to optimize distinction of the multiple groups. In our potato-projection example, projection of the two populations on the floor clearly separates the groups. In an actual DFA, the floor would represent the discriminant function. This function is an equation that can, in this case, use the x- and y-position values for an observation (potato) to determine to which population the object belongs. Using a distance-to-population-center measurement in n space, DFA also will calculate the probability (degree to which) that the observation belongs to the assigned population. This probability can essentially be thought of as a measure of how far the object "plots" in n space from the gravimetric center of the population. DFA calls this probability a "posterior probability." In the case of our doorframe, the discriminant equation would be a simple equation for a line. In most DFAs, equations are complex and represent convolute surfaces in n space that separate populations.

Posterior probability values are risk assessment tools. For example, if a discriminant function is established for our potato problem and then a 101st potato is evaluated by the function, DFA might assign the new observation to the sack A population with a posterior probability value of 0.2. Posterior probabilities generally range from 0 to 1.0, respectively indicating a "weak" and "strong" confidence of "belonging" to the assigned population. In this example, the new potato was assigned to the sack A population. The 0.2 posterior probability (20% probability) indicates a relatively weak relationship between the newly evaluated potato and the majority of potatoes that constitute the sack A population (the new potato plots relatively far from the gravimetric center of the sack A population). In this and other ways, DFA can be used as a powerful risk assessment tool to help determine the probabilities associated with a wide range of events.

BAYESIAN ANALYSIS

Bayesian analysis facilitates the use of new information to update (modify) initial probability estimates. It also can use historical probabilities to revise the probability estimates associated with a new project or process. It is a powerful risk assessment and management tool. Bayesian analysis generally requires that each component of a project or process have an associated (estimated, at least) probability (chance of happening).

Bayes' theorem and associated equations were devised by the 18th-century English clergyman and philosopher Thomas Bayes as part of his study of inductive reasoning and logic. It primarily is used to analyze the probabilities associated with the variables that comprise any process or project.

Most projects and processes conclude in a result. When the project or process is completed, the final result is obvious. This final state, however, generally is not so obvious at the onset of the work. That is, at the beginning of the project, there usually are several or many potential or probable outcomes (i.e., a project may succeed or fail, the introduction of a new product may result in a wide range of possible sales scenarios, etc.).

At the beginning of a project or process, the initiators of the work may assign probabilities to each step of the work and a total probability can be calculated for each of the many possible outcomes. A simple example might be the environmental remediation of a drilling site. Prior to the remediation effort, there are many variables to consider which constitute the remediation task. Some of these might be the issuing of permits by state regulators, the cost of sample collecting, the cost of sample analyses, the purchase of special equipment if the analyses show abnormally high concentrations of toxins, and other factors. Each of the states of these variables may be assigned a probability. Probabilities assigned before the actual start of a process or project are called "prior probabilities."

For business reasons, the company responsible for the remediation may be interested in the total probability of two states. These might be whether or not the cost of the remediation is going to be more or less than $250,000. To determine the probability of the remediation costing more or less than $250K, the company can assign dollar values and probabilities to each possible state of the variables. For example, the total cost of the project without considering sampling costs is $200,000. It may be determined that there is a 60% chance (probability = 0.6) that the state will not issue the permits, given the present remediation proposal. If the state does not issue the permits, it will cost an extra $25,000 to modify the proposal. Based

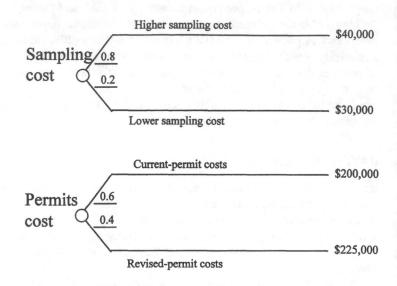

Total cost = Sampling cost + Permits cost =
((0.8 × $40,000) + (0.2 × $30,000)) + ((0.6 × $200,000) + (0.4 × $225,000)) =
($32,000 + $6,000) + ($120,000 + $90,000) =
$248,000

FIGURE 8.1   Example of combining probabilities and consequences.

on experience with the selected contractor, there is an 80% chance (0.8 probability) that the sampling will cost about $40,000. Similarly, probabilities and costs can be assigned to all other pertinent variables. We can calculate all relevant combinations of probabilities and dollar values to arrive at a probable cost. In Figure 8.1 is shown graphical example of this process.

Bayes' theorem also allows us to modify or update the prior probabilities as we learn new information during the course of a project or process. The actual equations associated with Bayes' theorem can be found in nearly any statistics book and will not be addressed here. Suffice it to say that the equations simply facilitate the integration of new information (new probabilities) that modifies the prior probabilities associated with the variables and with the various final-outcome states. In the example given here concerning a drilling-site remediation, the company may learn that new legislation is pending in the state legislature that has an associated probability of significantly altering the permit issuance process. Bayes' equations facilitate the introduction of the new information and the updating of the probabilities associated with the various end states. This is a powerful business and planning tool.

An associated use of Bayes' work is to update the probability attached to a "new" object based on the prior probabilities associated with similar historical objects for which the outcomes have been determined. An example can be found in the pre-drill estimate of the total chance of success associated with oil and gas prospects.

A given company may be employing a prospect-evaluation risk model that is composed of nine variables. Each variable may have an associated chance of failure (see Chapter 11 for a full treatment of the chance-of-failure concept). The nine chance-of-failure values are combined to give a total chance of success for a prospect. The company presently is using the nine-parameter risk model to evaluate a new prospect in the North Sea for which they have estimated chances of failure for each of the nine variables and, in turn, a total chance of success for the prospect. However, they now wish to use Bayesian analysis and historical data to validate the prospect's calculated chance of success.

In the company's database they have analogous North Sea wells that have been drilled. These drilled wells represent both successful wells and dry holes. A major subset of these wells will be those for which their pre-drill estimates of the nine chances of failure and the total chance of success were representative of what was the eventual outcome for each well (i.e., their pre-drill estimates were on target). Baysian analysis provides them a means to use this subset of historical and analogous prospect pre-drill information to "update" or modify the new well's chances-of-failure estimates and the new well's total chance of success. This is a powerful risk assessment tool that can use analogies to predetermine the probabilities associated with a new project.

A third, but by no means final, application of the work of Bayes is related to decision-tree analysis. This application of Bayesian-like analysis facilitates the classification and characterization of objects upon which we have measured the values of several variables.

Let's consider a sludge-pit example. On five pits we may have measured pit area, the pH of the fluid in the pit, and the percent clay in the material lining the pit bottom. Two of the five pits have developed leaks. A decision-tree-like plot can be constructed to characterize (hopefully) those pits that leaked and those that have not. Such a diagram is shown in Figure 8.2.

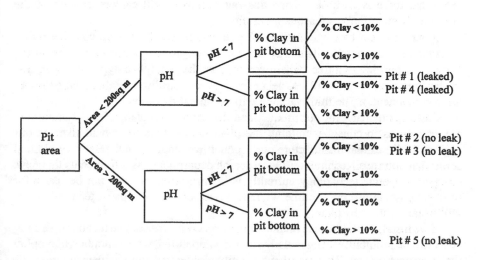

FIGURE 8.2   Decision tree for the purpose of discriminating between pits that have leaked and those that have not.

In Figure 8.2, leaking and nonleaking pits have been segregated and character-ized. Probabilities could be substituted for the numerical values in Figure 8.2 so that Baysian methods could be used to calculate the probabilities associated with each of the end states. Diagrams (decision trees) like these also can be used to determine, in this case, whether a "new" sixth pit is likely to leak or not.

DECISION TREES

In the immediately preceding section I defined just how a fundamental decision tree is constructed and illuminated the salient qualities of decision trees in general. I will, in the immediately following paragraphs, describe some inherent limitations in their application.

Decision trees are the near-perfect vehicle for graphically describing a series of decisions that will result in a course of action. However, decision trees are not a panacea when applied to analytical processes related to the same problem.

For example, we may be charged with the responsibility for deciding whether to build a new plant (factory). Our initial attempt to arrive at a decision will be a rudimentary one including only the following variables:

- Customer base (CB) in units of MM people;
- Percent of customer base captured (PC) in units of percentage;
- Operating costs (OC) in units of $MM.

In our decision tree we will consider the variables in the order in which they appear in the list above. Therefore, we will first consider whether there is a sufficient customer base. Demographic studies indicate that there is a 90% chance that the customer base is of sufficient magnitude. Similar studies by consultants have indi-cated that there is an 80% chance that our company will capture enough of the customer base to make a reasonable profit.

If our customer-base calculations have favorable results, we will consider build-ing a plant. In our admittedly simple example here, we will consider only one more variable in our decision tree (when we consider the "analytical-type" decision tree in the following text, it will be clear why we are here limiting this study to only three components). The third variable will be operating costs.

Customer demand for our product and the announced plans of our competitors are causing us to consider building the plant sooner than we would like. At our existing facilities that manufacture our product, we employ some relatively old and somewhat outmoded technology. Our research department has informed us that new, cheaper-to-operate technology currently is being tested, but may not be ready for production-scale use by the time we have to decide on the type of technology we will install at the new plant.

If we install our traditional manufacturing process, we estimate that our operating costs will be prohibitively high relative to our competition, who no doubt will employ the latest technology. We might have to utilize older and more expensive methods at the new plant in order to beat the competition to the marketplace. Therefore, our experts estimate that we likely have about a 60% chance that operating costs at the new plant will be prohibitively high.

I consider there to be two basic types of decision trees (others may disagree — that is their prerogative) that commonly are built. These I will term the "conceptual tree" and the "analytical tree." To construct a conceptual tree, only probabilities and single deterministic leaf-node (extreme "right hand" end of any branch) values are required. A typical conceptual tree for the problem outlined above appears in Figure 8.3.

In Figure 8.3, the aforementioned probabilities are indicated on the appropriate branch. So, in the case of customer base, there is a 90% chance that the customer base will be sufficient to sustain our business. Conversely, there exists a 10% chance that the base will be too lean and our enterprise will fail.

Considering the percentage of the customer base we are likely to capture, our consultants have estimated that there exists an 80% chance that, if we move quickly, we will capture a sufficient share of customers to make our new plant a successful economic venture. The 20% chance of capturing an insufficient number of customers reflects the facts that the customers, for whatever reason, might largely shun our product and/or the probability that the competition might beat us to the marketplace. As already explained, we believe there is a significant chance (60%) that our operating costs will exceed our acceptable limits. This is duly noted on the decision tree.

At the far right-hand end of the tree is a leaf node to which a consequence has been assigned. In this case, the consequence value is the deterministic amount of money (in millions of dollars) we expect a successful plant to generate.

Back at corporate headquarters, it has been decided that, to be economic, the plant must generate at least $10MM in income per year. To determine whether or not this enterprise is capable of infusing the corporate coffers with sufficient funds, the probabilities and leaf-node value are combined in a deterministic equation thus:

$$\text{Probable income} = 0.9 \times 0.8 \times 0.4 \times \$40\text{MM} = \$11.52\text{MM} \qquad (8.1)$$

This calculation results in a value of $11.52MM, indicating that our enterprise likely will exceed the $10MM corporate-headquarters-imposed economic hurdle. Such calculations give us some concept concerning whether, in this case, the proposed plant will be economically successful. Decision trees of this type can be invaluable in graphically outlining a problem and in determining, on a conceptual level, the likelihood that a proposed project will succeed or fail.

Decision trees, however, seem to reach their practical limit with the conceptual model. That is, when copious quantities of data, complex probabilistic branching, time-series analysis, etc., are introduced, the decision-tree approach becomes, at least, unwieldy. I use the term "analytical" (many other terms are used by myriad people to describe the same thing) to indicate a decision tree into which such information has been injected.

Consider, for example, the plant-building situation several months after the construction of the conceptual decision tree. Over the past few months, teams of experts in the fields of demographics, marketing, and operations have individually met to "put some numbers" to the conceptual model. In the "old business school"

CB = Customer Base

PC = Percent Customer Base captured

OC = Operating Costs

STOP = Failure of project due to unacceptable values for one of the components

⌐40 = Estimated yearly profit from project

10 = Probability of a given branch

FIGURE 8.3 Example of a "conceptual" decision tree.

style of assigning ranges to variables, management has requested that each team come up with a reasonable maximum value, a mean or "medium" value, and a low value. The low-values branch for customer base and percent-of-customer-base captured on the analytical decision tree is to represent the chance of attaining a "minimum success" value. High values for operating costs are to have similar meaning.

In addition, each team is to fabricate a "leaf-node" value for each of the branches. To generate such values, representatives from each team have to meet to discuss the scenario represented by each branch. Again, in the "old business-school" fashion, the teams are required to generate but a single value for each leaf node (a bit later in this section, I will discuss the use of distributions in conjunction with decision trees).

In Figure 8.4 is shown the analytical decision tree for this project. I mentioned when describing the conceptual decision tree above, that the reason for limiting this example to just three variables would become clear when an analytical tree was built for the project. It should be painfully obvious that this methodology is not an expandable architecture. The tree with three variables already has 27 leaf nodes to which values have to be ascribed. Addition of just one more variable raises the number of leaf nodes to 81.

I have performed numerous risk assessments for projects like the one described here. Considering technical, scientific, engineering, commercial, environmental, financial, political, and other component parts of such a risk model, it is not uncommon for this type of model to be composed of 50 or more variables. I have found it to be impractical to use the decision-tree type of analysis on real-world risk models. A decision tree turns into a "decision bush."

In addition to the proliferation of branches, I find it simply unreasonable to expect anyone or any team to be capable of (or have the time for) talking through the logic of each branch and ascribing a value to each leaf node. For the tree shown in Figure 8.4, teams would be required to talk through 27 logic trains. Addition of a fourth variable commits them to discuss 81 separate scenarios.

Aside from the node-proliferation problem, most real-world risk assessments of this type consist of components and processes that are not well represented by the analytical-decision-tree approach. Severely branching logic (i.e., "IF" and "ELSE" statements), looping (i.e., "WHILE" statements), time-series analysis (considering the third dimension of time), and other commonly employed risk-model-building techniques I find to be all but excluded by the analytical-decision-tree approach.

Solving the tree is elementary. A decision-tree practitioner need only sum the products of each branch. The tree shown in Figure 8.4 would be solved by the following system of equations.

$$
\begin{aligned}
\text{Plant profit} = &\ .1 \times .2 \times .1 \times \$7\text{MM} \\
+ &\ .1 \times .2 \times .3 \times \$6\text{MM} \\
+ &\ .1 \times .2 \times .6 \times \$5\text{MM} \\
+ &\ .1 \times .7 \times .1 \times \$8\text{MM} \\
+ &\ .1 \times .7 \times .3 \times \$7\text{MM} \\
+ &\ .1 \times .7 \times .6 \times \$6\text{MM} \\
+ &\ .1 \times .1 \times .1 \times \$9\text{MM}
\end{aligned}
\tag{8.2}
$$

FIGURE 8.4   Three-variable "analytical" decision tree.

$$+ .1 \times .1 \times .3 \times \$8MM$$
$$+ .1 \times .1 \times .6 \times \$7MM$$
$$+ .6 \times .2 \times .1 \times \$10MM$$
$$+ .6 \times .2 \times .3 \times \$9MM$$
$$+ .6 \times .2 \times .6 \times \$8MM$$
$$+ .6 \times .7 \times .1 \times \$14MM$$
$$+ .6 \times .7 \times .3 \times \$12MM$$
$$+ .6 \times .7 \times .6 \times \$10MM$$

$$+ .6 \times .1 \times .1 \times \$15MM$$
$$+ .6 \times .1 \times .3 \times \$13MM$$
$$+ .6 \times .1 \times .6 \times \$12MM$$
$$+ .3 \times .2 \times .1 \times \$18MM$$
$$+ .3 \times .2 \times .3 \times \$16MM$$
$$+ .3 \times .2 \times .6 \times \$15MM$$
$$+ .3 \times .7 \times .1 \times \$19MM$$
$$+ .3 \times .7 \times .3 \times \$23MM$$
$$+ .3 \times .7 \times .6 \times \$16MM$$
$$+ .3 \times .1 \times .1 \times \$20MM$$
$$+ .3 \times .1 \times .3 \times \$18MM$$
$$+ .3 \times .1 \times .6 \times \$17MM = \$12.049MM$$

The tree's resultant value of $12.049MM exceeds the hurdle value of $10MM set by corporate headquarters. In fact, it exceeds the conceptual-tree estimate of $11.52MM. So, the project likely will go ahead.

Some of the sting of employing analytical decision trees can be abated by the use of modern software packages that allow the user to represent each leaf node with a complete distribution. This technique may or may not reduce the number of branches and nodes (depending on how it is employed). It does, however, avoid the problem of "discretizing" (selecting a small and fixed number of values to represent an entire distribution) which, given modern software and statistics-free distribution-building technology, is never necessary or even desirable.

Another practical approach to addressing analytical decision trees that have grown into decision bushes is to simply convert them to a Monte Carlo-type model. This approach will sacrifice the graphical "tree" representation of the problem (the graphical tree representation can hardly be considered an advantage in a many-variable tree) but does afford many advantages. Using entire distributions instead of a limited number of discrete values is facilitated. This process allows the model to retain all of the relevant data. In addition, functions such as time-series analysis, branching, looping, and other processes are easily accommodated. See the example Decision-Tree Conversion in Chapter 14 for an illustration of such a conversion.

FACTOR ANALYSIS

Of the technologies presented here, factor analysis is the most difficult for people to conceptualize and understand. This, of course, means that it is difficult to explain. However, I am obligated to take a whack at it, so here goes.

The fundamental purpose of factor analysis is to reduce the dimensions of a problem. This means that factor analysis is good at reducing the dimensionality of the variable space in which data points are plotted. For example, we may have accessed the census bureau files on a group of people. The files contain a large number of variables (let's say 100) measured on each person. If you were to plot each person in 100-dimensional space and could view the plot, you might see some grouping of people (clustering, if you will) but would be hard pressed to tell anyone which variables contributed significantly to the clustering of bodies. Factor analysis helps you to reduce the dimension space (reduce from 100 to some manageable number) by creating a new set of uncorrelated variables that contain "most" of the information contained in the original set of 100 variables.

Geometrically speaking, uncorrelated variables and their axes are those, in a 3-D world, that when plotted have axes at right angles to one another. One of the mystifying aspects of statistics is that you can have as many axes at right angles (uncorrelated) as you like. For us 3-D thinkers, it bends our gray matter all out of shape to try to imagine such a thing. Factor analysis is steeped in jargon. Some of the more esoteric terms are "eigenvector" and "eigenvalue." These relate to the uncorrelated axes discussed in the previous paragraph. Let me explain how.

Imagine a cloud (like you see in the sky) floating in space. We can only imagine three-space (not counting time), but remember that in the statistical world, this "cloud" could be "floating" in n-space. All the particles which comprise the cloud can be thought of as data points plotted in n-space. Let us assume that n is some unmanageable number of variables, and we would like to "boil down" the number of variables to a smaller, more comprehensive, and representative set of parameters.

Imagine now that you are standing inside the cloud and have with you a measuring tape. This tape can get longer, but cannot bend, so you can only measure distances along straight lines (in three-space). What you would like to know is, how long is the longest straight line you can draw within the confines of the cloud? Through a process that I will not describe here, factor analysis facilitates the discovery of this longest line through the cloud. A line such as this is called an "eigenvector" and its "length" is called the "eigenvalue" for that eigenvector. This, loosely, can be considered the "first principal component" in factor-analysis talk.

This longest eigenvector represents the "direction" through the cloud that describes the greatest amount of variability in the "shape" of the cloud. It describes more variance in the plotted data than any other line you can draw. This line or eigenvector is represented by an equation, the terms of which are all the original variables.

The first eigenvector may account for, say, 55% of the variance of the dataset. Because we want a new set of variables that describe most of the variance represented by the original set (n) of variables, we should not be satisfied with 55% and will seek to find a second eigenvector.

The second eigenvector is the second-longest line you can draw through the cloud, with the "catch" that this second line must be at "right angles" (again, the term "right angles" only is good in a 2- or 3-D world) to the first eigenvector; i.e., it is uncorrelated. Again, factor analysis facilitates finding this axis, which describes the second greatest amount of variance. This axis may account for, say, 30% of the variance in the dataset. Now we have accounted for 85% of the variance. The identification of about two more eigenvectors will likely bring us to a percentage variance value greater than 95. In the world of factor analysis, the contributions of additional eigenvectors becomes increasingly small, so it rarely pays to define more than the number of vectors that it takes to describe 95% to 99% of the variance.

When factor analysis is through finding axes through the cloud, we have a much reduced number of variables that do a good job of describing what this cloud looks like (describing the data distribution). Well, what in the world does this have to do with risk assessment?

Each original variable will be a term in the equation of each eigenvector. Let us say that we have a cloud floating in three-space (X, Y, Z) and that its longest dimension is nearly parallel to one of the original axes — let us say the X axis. It stands to reason that the longest line that one can draw through the cloud will be nearly parallel to X. This means that X will be a very "important" or "influential" variable in the equation that describes the nearly X-parallel eigenvector. The measure of how "important" a variable is to the equation that describes the eigenvector (how parallel the variable axis is to the eigenvector) is called a "loading" for the variable. Each original variable has a loading value for an eigenvector. These loadings can be used to determine which variables are the "most important" in a given dataset. This can be very significant information when you are attempting to perform a risk analysis but you are faced with a large set of variables, and you have no idea as to which ones you should pay most attention. The loading values can be thought of as "weightings" for the variables for use in risk-weighted calculations. This kind of analysis can be exceedingly helpful when attempting to determine for a risk analysis which are the variables that are most likely to "bite" you. The risk analysis, however, may actually be run on the newly defined (and smaller number of) "factors."

Now that I have played up this technology, let me deflate it a bit by mentioning an associated practical problem. Eigenvectors are great things and are powerful data analysis tools, but when you use them as the variables in an analysis and your boss asks you what they mean (i.e., "Glenn, what is the significance of factor 2 having such a high value in the final analysis?"), you had better be prepared to describe just what is factor 2. This often is difficult because factor 2, like all other factors, is a combination of all the original variables. If you are lucky, factor 2 will have high loadings for only one or two of the original variables. However, it often is the case that many original variables have significant contributions to factor 2 (have high loadings for factor 2), and it becomes practically impossible to describe to the uninitiated just what is the meaning of a change in factor 2. This explanation becomes even more difficult when the relative changes of multiple factors have to be explained. What this means in a practical sense is that factor analysis is a very powerful data analysis and risk assessment tool, but is not much of a "presentation" vehicle.

NEURAL NETS

As the name would imply, neural nets are computer systems that try to emulate the function of the brain. Neural nets (NNs) can be thought of as a branch of the more general sciences of artificial intelligence and expert systems. NNs, however, have special characteristics that make them not only unique, but fundamentally more powerful and generally applicable than many other artificial intelligence techniques or expert systems.

As already stated, the aim of NNs is to build a computer system that can process information like the brain. The fundamental elements of a brain are, to oversimplify the situation, the neuron, synapses, dendrites, and axions. Dendrites collect information from other brain cells and from the "outside" and communicate this chemical message through synapses to the neuron. The neuron is the basic cell in the nervous system that receives and combines information. Output from a neuron is carried to other cells, etc., by the axions. The human brain is composed of billions of these structures.

The fundamental elements of a typical NN are designed to emulate these basic nervous system elements. The "processing element" in an NN is the (rudimentary) equivalent of the brain's neuron. The processing element receives information from many input paths (dendrites and synapses) and internally combines the inputs and "makes decisions" about passing out output signals.

The information carried by an input path to a processing element usually is weighted. Multiple weighted "pieces" of information are summed in the processing element. The processing element also contains a "transfer function." This function can be a simple threshold (i.e., if the sum of the weighted input information is not great enough, then no output is generated) or can be a complex, continuous function of the summed input. The output from the transfer function is passed to the output path (axion) which, in turn, can be used as input to subsequent processing elements. NNs are comprised of many such connections of processing elements. Typically, the processing elements are configured in layers.

One of the real advantages of NNs is their ability to "learn." The learning process is fundamentally a matter of, for a given input, generating the desired output by adjusting the weighted input at the various processing elements. There are many types of learning (supervised, unsupervised, reinforcement, etc.), the details of which will not be conveyed here. Suffice it to say, NNs offer the ability to generate a desired response and to modify the response through time (and "experience") without having to write a traditional computer program for each unique desired outcome.

NNs are excellent at such things as pattern recognition, signal processing, target classification, and many other tasks. One of the tasks to which NNs can be put is that of prediction. It has been shown that NNs, in certain cases, offer orders of magnitude improvement over more conventional linear and polynomial methods of predicting. Much of the use of risk assessment (Monte Carlo analysis, etc.) is involved with predicting the possible outcomes of a situation and the likelihoods associated with each probable outcome. In appropriate situations, NNs offer a powerful, practical, and flexible method of risk assessment.

Another area where NNs may become practical in the future is that of system modeling. Many of the risk assessments performed are of the "process" type. That is, they are not basically probabilistic in nature but rather attempt to emulate a natural process in order to evaluate the possible outcomes. Although probably not quite sophisticated enough yet to handle complex process models, NNs likely offer a flexible solution to relatively simple process situations.

9 Monte Carlo Analysis

CONTENTS

A Bit of History..91
For What Is It Good? 92
Simple Monte Carlo Example...92
How Many Random Comparisons Are Enough?..94
Output From Monte Carlo Analysis — The Frequency and Cumulative
Frequency Plots 95
Interpreting Cumulative Frequency Curves 100
Combining Monte Carlo Output Curves... 104

A BIT OF HISTORY

Monte Carlo analysis uses the process of simulation to achieve a range of solutions to a problem. This technique generally is used to solve problems for which the definition of specific solution equations to "calculate" a specific answer is either too complex or too cumbersome to be practical. In Monte Carlo analysis, typically (but not necessarily) input variables are represented as distributions of values (see Chapter 10 of this book). Values comprising these distributions represent the range and frequency of occurrence of the possible values for the variables. Also needed are equations that relate the input variables to the output (result or answer) variables.

Monte Carlo analysis is a statistical procedure based upon "random" selection or chance. The name "Monte Carlo," of course, is taken from the city of Monte Carlo, Monaco, made famous by games of chance.

Monte Carlo analysis, or distribution sampling, arguably first was presented formally by "Student" (W. S. Gosset, inventor of the Student's t-test) in the early 1900s. At first, Monte Carlo analysis was treated as being distinct from "experimental sampling," or "model sampling," both of which involve random selection from populations. The term "Monte Carlo" was restricted to processes in which weighted or importance sampling was used to increase the efficiency of the sampling process. In other words, the term "Monte Carlo" was reserved for sampling processes that embodied some "smarts" to increase sampling efficiency. This restricted use of the term eventually faded because high-speed computers lessened the importance of an efficient sampling process.

Today, the term "Monte Carlo analysis" can be applied to any procedure that uses distribution-based random sampling to approximate solutions to probabilistic or deterministic problems. These types of problems were exemplified by the first practical use of the Monte Carlo method — addressing the probabilistic aspects of particle diffusion associated with the design and construction of the first atomic weapons at Los Alamos during World War II.

FOR WHAT IS IT GOOD?

Most processes, physical, chemical, or otherwise, terminate in a result. Many processes are so complex that the generation of an equation or system of equations to calculate the result of the process is arduous or impossible. Monte Carlo analysis often is used to simulate these processes and to generate probable results. This avoids having to resort to the construction of specific equations which can predict the outcome of the process. In addition, Monte Carlo analysis can yield likelihoods for a range of possible results.

Although Monte Carlo analysis has many uses in business and science, the most common application is that of determining the probability that a certain event (or result) will occur and getting a grip on the magnitude of the event. The fundamental technique used to accomplish these things is the random sampling of values from frequency distributions which represent the variables in the analysis (see Figures 9.1 and 9.2). Well, what does all that mean?

Every risk assessment study involves the integration of at least several and sometimes many variables. Variables are the basic elements of the problem (porosity of the soil, concentration of an element in groundwater, distance from a potable water source, etc.). If Monte Carlo analysis is to be used, each of these variables usually is represented by a distribution of values (coefficients of the variable). The values that constitute the distribution should span the range of possible values for the variable in question.

Monte Carlo analysis, in its native form, also requires that the variables be independent. That is, they must not have a relatively high correlation. For example, if two variables in a Monte Carlo analysis are depth and temperature, then it is not valid to randomly select a depth and temperature value from their respective distributions and expect that this combination will emulate a real situation. This is because, in general, as depth increases, temperature increases. A randomly selected depth of 5 feet and a temperature of 500°F do not, in most cases, simulate real-world conditions. To account for the correlation between variables, the technique of defining dependence between the two variables should be employed. See Chapter 12 of this book for an explanation of this technology.

SIMPLE MONTE CARLO EXAMPLE

For the sake of an example of Monte Carlo analysis, let us consider the oversimplified example of calculating profit from income and payments (payments = salaries, taxes, and other costs). Figure 9.1 shows a distribution of possible incomes for the year. Figure 9.2 shows the range of payments we make in the same year. On each iteration of Monte Carlo analysis, a randomly selected income will be combined with a randomly selected payment to yield a profit value using, for example, the following equation:

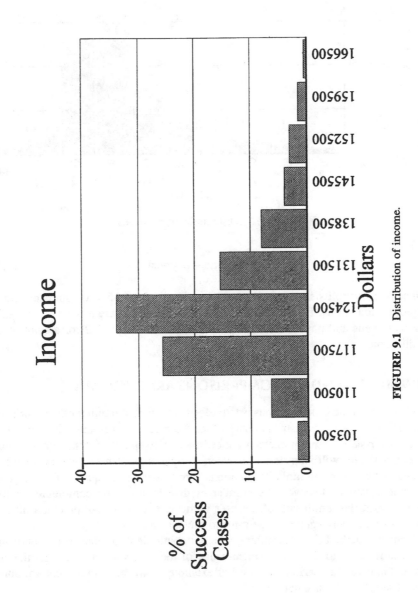

FIGURE 9.1 Distribution of income.

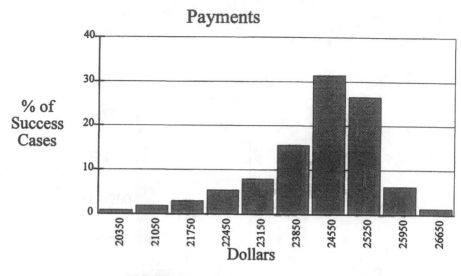

FIGURE 9.2   Distribution of payments.

$$profit = income - payment \qquad (9.1)$$

The plot of profits shown in Figure 9.3 is the result of 1500 random selections and profit calculations. This model, of course, assumes that there is not a relationship between income and payments. This is not likely. See Chapter 12 for a solution to this dilemma.

HOW MANY RANDOM COMPARISONS ARE ENOUGH?

The object in Monte Carlo analysis is to perform a sufficient number of comparisons from the input variable distributions so that the resulting variable distribution is valid. Some risk assessment packages allow the user to specify a convergence value that automatically will stop iterations when comparisons of mean values (or other selected values) from multiple "simulations" are less than or equal to the assigned convergence value. However, the number of distributions, the complexity of the equations, and the familiarity of the user with the situation may preclude his/her knowing such a convergence value prior to the analysis.

A simple method for determining whether a sufficient number of comparisons have been made is to inspect the mean value of the "answer variable" distribution from multiple Monte Carlo analyses. For example, if we have two input variables called X and Y and our equation is

$$Z = X + Y \qquad (9.2)$$

then the Z distributions from multiple individual Monte Carlo analyses (an individual Monte Carlo analysis may be comprised of, say, 5000 iterations) would be inspected.

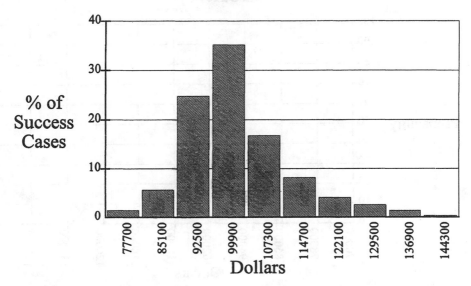

FIGURE 9.3   Distribution of profit resulting from Monte Carlo combination of income and payments.

If the means of the individual resulting distributions do not vary significantly (significance is determined by the user), then it is likely that a sufficient number of comparisons has been made. Statistical tests for the significance of the difference of population means (t-tests) or of the entire distribution using the Kolmogorov–Smirnov test are available but can be complicated. Most of the time, simple inspection of the change in the mean of the resultant distributions and the shape of the distributions will suffice.

OUTPUT FROM MONTE CARLO ANALYSIS — THE FREQUENCY AND CUMULATIVE FREQUENCY PLOTS

As depicted in Figure 9.3, the primary output vehicle for Monte Carlo analysis is the frequency plot. The number of bars in a frequency plot can be set by the person generating the model. Bar heights simply indicate the number (or percent) of X-axis values that fall within the X-axis range dictated by the width of the bars.

Although useful, frequency plots are limited in the amount and type of information conveyed. In risk assessment and portfolio management, it is more common to view Monte Carlo output in cumulative frequency space. The same data shown in frequency space in Figure 9.3 is depicted in cumulative frequency space in Figure 9.4.

The conversion of frequency space to cumulative frequency space can be conceptually viewed as "stacking" the bars of a frequency plot. Figure 9.5 shows a four-bar frequency plot. In Figure 9.6, the same four bars are "stacked" from right to left. In Figure 9.7, a dot has been inserted at the top of each bar and the dots connected

Profit

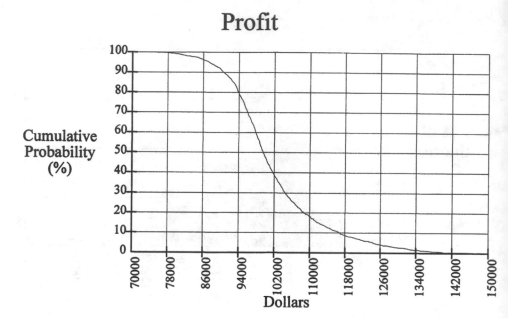

FIGURE 9.4   Cumulative frequency plot equivalent to frequency plot in Figure 9.3.

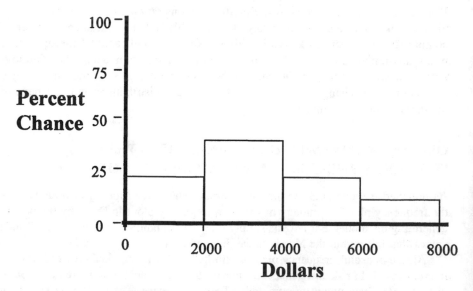

FIGURE 9.5   Four-bar frequency plot.

with a line. In Figure 9.8, the bars have been removed, leaving just the cumulative frequency curve.

In the computer, of course, there actually are no bars. Software systems that build cumulative frequency curves work directly with the individual X-axis values

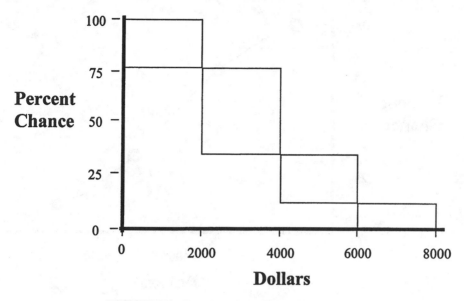

FIGURE 9.6   Four bars from Figure 9.5 stacked.

FIGURE 9.7   Curve connecting tops of bars from Figure 9.6.

generated by Monte Carlo analysis. In Figure 9.9 is shown a plot of eight values resulting from an eight-iteration Monte Carlo risk assessment model. The eight values have been sorted from "smallest" (values of least magnitude) on the left to "largest" (values of greatest magnitude) on the right.

FIGURE 9.8   Cumulative frequency plot equivalent to frequency plot in Figure 9.5.

Cumulative Frequency Curve - 8-Iteration Case

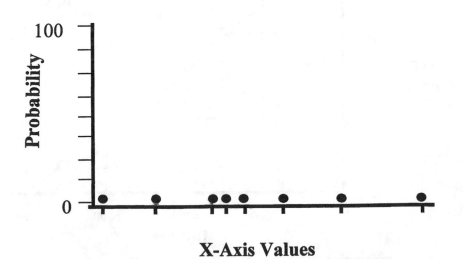

FIGURE 9.9   Results from an eight-iteration Monte Carlo analysis.

Because eight values were generated, the cumulative frequency algorithm will divide the Y axis into eight equal probability segments. The "largest" value is moved up one probability segment. The second largest value is moved up two probability

Cumulative Frequency Curve - 8-Iteration Case

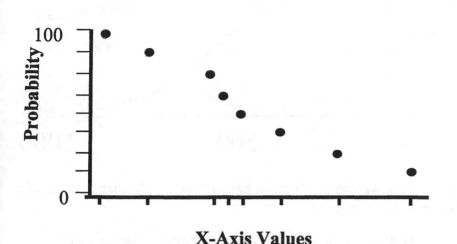

FIGURE 9.10   Individual X-axis values "raised" parallel to the Y axis according to their cumulative probability.

Cumulative Frequency Curve - 8-Iteration Case

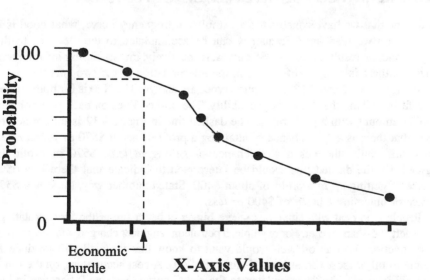

FIGURE 9.11   A cumulative frequency curve constructed by connecting the points in Figure 9.10.

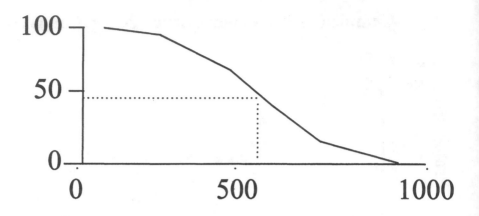

In this case, we can say that we have about a 45% chance of getting "at least" $550 or more.

FIGURE 9.12   Example of a "Z-shaped" cumulative frequency curve.

segments and so on until the algorithm reaches the last data point (leftmost or "smallest" X-axis value), which is moved up, in this case, eight probability segments. The points are connected with a smooth line which is the cumulative frequency curve (Figures 9.10 and 9.11).

INTERPRETING CUMULATIVE FREQUENCY PLOTS

Well, now that we have constructed a cumulative frequency curve, what good is it and how do we read one? Frequency can be accumulated to the "left" or to the "right," and the resulting "Z" or "S" curves, respectively, impart similar information. However, there is a slight difference in the interpretation of the two curves.

In Figures 9.12 and 9.13, both curve types are shown. The X axis for both curves is profit in dollars; the Y axis is probability. The dashed lines on each plot connect a dollar amount with a probability. The dashed line in Figure 9.12 is interpreted to mean that there is a 45% chance of attaining a profit of about $570 *or greater.* Or, stated differently, there is a 45% chance of getting *at least* $570 of profit. In Figure 9.13, the dashed line would be interpreted to indicate that there is a 65% chance of getting *up to* a profit of about $400. Stated another way, there is a 65% chance of obtaining a profit of $400 *or less.*

People who deal with situations where bigger is better prefer the "accumulating-to-the-left" (Z-shaped) plot. For example, a petroleum engineer charged with estimating oil production from an oil well would want to know that the well will produce X amount of oil or more (or, at least X amount of oil). A cost accountant, on the other hand, might prefer the "accumulated-to-the-right" curve (S-shaped) because in her job it is important to know that a project has a Y-percent chance of costing "up to" X-many dollars.

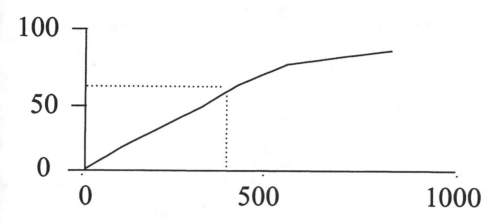

In this case, we can say that we have about a 65% chance of getting "up to" 400 (including 0).

FIGURE 9.13 Example of an "S-shaped" cumulative frequency curve.

Most off-the-shelf risk-assessment packages offer the S curve as the default situation. However, those with experience in all types of risk assessment prefer the Z curve because its "high-probability" end is the end near the Y axis. This has great advantage when incorporating chance of failure into a risk assessment (see Chapter 11 of this book) or when calculating risk-weighted values (see Chapter 13), or when performing a host of other risk assessment tasks. Therefore, I recommend the Z curve as the most utilitarian projection of cumulative probability.

A typical misconception associated with cumulative frequency plots is that the "low-value" end of the curve (the end nearest the Y axis) should approach or reach zero. Consider the situation of two distributions, A and B, displayed as arrays of values in Figure 9.14. In each distribution, the smallest value is 2, and the value 2 only occurs one time in each distribution. The risk model that uses distributions A and B is simply using Monte Carlo random sampling to add A and B to produce C (C = A + B).

The value of least magnitude that C can attain is 4 (A = 2 and B = 2). Given that there are 20 values each in the A and B distributions, we have a 1 in 400 (20 × 20) chance of producing a value of 4. You will note, however, that the value of 4 plots at 100% on the Y axis in Figure 9.15. It should be apparent that there is not a 100% chance of producing a value of 4 by random selection from A and B (only 1 chance in 400). So, what do we have a 100% chance of producing with the equation C = A + B? There is a 100% chance of calculating a C-value of 4 *or greater*. Other X- and-Y-axes pairs are similarly interpreted (a Y probability of getting X or greater than X).

This leads to another point concerning the interpretation of cumulative frequency curves. In Figure 9.15, the value of least magnitude on the plot is also the value of least magnitude that could have been produced from the distributions A and B. Similarly, the maximum value on the curve is 26 that also is the value of maximum

Variable A	Variable B
3	7
5	2
6	13
2	4
12	5
4	10
8	9
13	11
10	6
9	12
7	8
11	10
4	3
9	10
12	9
5	11
8	4
10	7
9	3
11	5

FIGURE 9.14   Two 20-value arrays for variables A and B.

magnitude that could be calculated from distributions A and B (given our C = A + B equation). It is not typical that the end points of a cumulative frequency curve represent the minimum and maximum values that could have been produced.

Distributions A and B in Figure 9.14 each are comprised of only 20 values. If 400 or more Monte Carlo iterations are used to randomly sample distributions A and B, then there would be a pretty good chance of producing the minimum and maximum values that could result from adding A and B (the greater the number of iterations, the better the chance of producing the minimum and maximum values). Distributions in actual risk-assessment models typically contain 100 to 500 or more values. Also typical is to have many more than two distributions in the risk assessment model. Consider a model composed of, say, two distributions, each containing 200 values. We would need to run tens of thousands of Monte Carlo iterations to have any reasonable chance of producing a cumulative frequency curve that had as

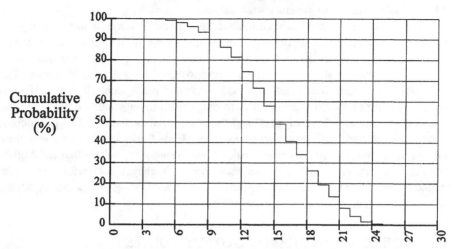

FIGURE 9.15 Cumulative frequency plot with a minimum value of 4 and a maximum value of 26. Curve created by summing variables A and B (Figure 9.14).

its end points the values of largest and smallest magnitude that could have been produced by combination of the two distributions.

Because of the common complexity of risk assessment models, the speed of the typical computer, and other factors, most risk assessment models are allowed to run a number of Monte Carlo iterations that is between 1000 and 10,000 (the number of iterations generally can be set by the user and can be highly variable). This is far too few iterations to ensure that values of minimum or maximum magnitude have been generated. Therefore, the "end points" of most cumulative frequency curves do *not* represent the values of minimum and maximum magnitude that could have been produced. The "end points" represent the values of minimum and maximum magnitude that were produced in the given number of iterations.

At first blush, this would appear a foible in the Monte Carlo process, but it is not. To illustrate the reason why you likely don't want absolute minimum and maximum values represented on the plot, consider the following situation. Suppose we are calculating profits from distributions representing number of units sold, price per unit, number of sales personnel, salary per hour for sales personnel, capital expenses, and operating expenses. If each of these distributions truly represents the range of values that could possibly occur (i.e., tails of the distributions are composed of values that are extreme and unlikely), there is a *very* small chance that, in the real world, there will be a simultaneous occurrence of the maximum number of units sold, the maximum price per unit, the minimum number of sales personnel, the minimum salary per hour for personnel, the minimum capital expense, and the minimum operating expense to produce the maximum possible profit. You may argue that it would be useful to calculate the minimum- and maximum-possible profit. I would counter this argument with the suggestion that you make such calculations on the back of a brown paper bag and that you don't need a risk-assessment or Monte Carlo analysis to generate such values.

The purpose of Monte Carlo simulation, among other things, is to produce plots of things that are likely to happen. Alignment of absolute minimums and maximums, in the real world, is not likely. It also can be dangerous to produce such values.

Consider the situation of a company composed of several sales locations. The company would like to calculate the total combined profit from all stores. If each of the store-profit curves were allowed to range from the absolute minimum to the absolute maximum value ("biggest" and "smallest" values that could be produced), summation of these individual curves, even by Monte Carlo analysis (the way they should be summed) will generate "upside" and "downside" values that are highly unlikely and extreme. These values may imbue "management" with unrealistic expectations (these numbers tend to get recorded in accounting sheets and reported). These numbers also can be downright scary.

COMBINING MONTE CARLO OUTPUT CURVES

In the example in the immediately preceding paragraphs, we considered the problem of combining (in this case, combining by summing) the Monte Carlo output profit curves from individual stores to arrive at a total-profit curve for all stores. It is tempting (and I see it done frequently) to add together all of the respective individual output curve minimums, means (or medians), and maximums. Another typical mistake is to sum (or otherwise combine) percentile values such as P10, P50, P90 values.

The reason that it is invalid to combine curves in this fashion is the same as that previously put forward regarding the combination of input distributions to Monte Carlo analysis and the calculation of values of absolute minimum and maximum magnitude. It is highly unlikely that extreme values will, in the real world, "align" to produce a resultant (sum) value of minimum or maximum magnitude. To correctly combine Monte Carlo output curves, each of the output curves should be used as an input curve to a new Monte Carlo routine specifically designed to combine the curves. When this approach is implemented, real-world and reasonable combinations and ranges result.

10 Decisions and Distributions

CONTENTS

Decisions 105
Just What Is a Distribution? 106
Distributions — How to Approach Them 110
Symmetrical Distributions 112
Skewed Distributions 114
Spike Distributions 115
Flat Distributions 116
Truncated Distributions 117
Discrete Distributions 119
Bimodal Distributions 120
Reading Data From a File 121
Peakedness 123
Triangular Distributions 125
"Specific" Distribution Types 126

DECISIONS

The only time we have to make decisions is when we don't know something about the situation. Consider the scenario in which your spouse asks you to go to the store. You leave the house and arrive at the proverbial fork in the road. You look down the right fork and see no store. You look down the left fork and you see the store. In this case, I submit that there is no decision to be made — the store is on the left fork and that is where you will go. However, if you reach the fork in the road and do not see a store down either fork, then a decision has to be made. Even if you decide to stay where you are or go back home, a decision still has to be made as to what to do. A decision has to be made because you don't know something. That is, you are uncertain about something related to the problem.

Distributions are one means of expressing uncertainty in a problem, and, in fact, using that uncertainty to our advantage. Therefore, distributions are used in decision making by affording us a means of quantifying, expressing, and using our uncertainty in the process of making a decision.

JUST WHAT IS A DISTRIBUTION?

One of the most common mechanisms for expressing uncertainty is the distribution. Just what does the term "distribution" mean? When you look at a distribution, just what should it be telling you?

Consider the situation of the new-hire engineer who has been summoned on his first day of employment into the office of the head of construction. The engineer is informed that as part of the plant-expansion construction, 2000 doors and door jambs will need to be built. The boss was informed by the supplier of materials that the company will have to buy its lumber in bulk directly from the mill. This means that the lumber will not be custom cut, but rather is available from the mill only in 1-foot increments. That is, the boards will come in 5-foot lengths or in 6-foot lengths or in 7-foot lengths, etc. The boss was unsure concerning what length lumber should be purchased, so she asked the engineer to measure the heights of several hundred adults in the community and present her with the measurement results so a lumber-length decision can be made.

Well, it is the engineer's first day on the job and he is anxious to impress the boss. In his zeal, he visited the local university and arranged to borrow some laser equipment they use to measure distances. The engineer set up the equipment to measure the heights of adults. The equipment is capable of measuring distances (heights) with great precision — out to 10 decimal places if needed.

After a week of collecting data, the engineer considered he had amassed a sufficient number of measurements and he produced a plot of the heights. A plot of a subset of the data appears in Figure 10.1.

What is a Distribution?

FIGURE 10.1   Measured values sorted and plotted on the X-axis.

What is a Distribution?

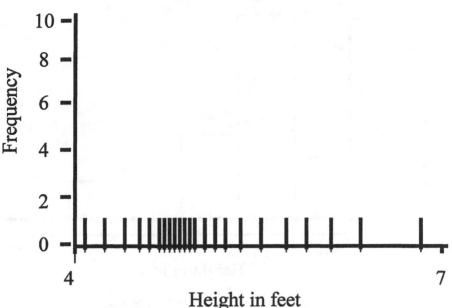

FIGURE 10.2 Plot of X-axis values each represented by a bar of height 1.

The measurements were carried out with sufficient precision that there are no repeating values; that is, each value is unique. So, the frequency with which each value occurs is 1. The plot of the data that the engineer took to the boss is shown in Figure 10.2.

The boss took one look at the plot and informed the engineer that the plot with which the engineer has presented her is not quite what she had in mind. She informed the engineer that she was expecting a more "typical" frequency plot with the bars of various height, and she sent him away to have another try.

"Well," the engineer wondered, "what is wrong with the plot I made?" After all, in a data set of nonrepeating values, each value occurs once (the probability of any value occurring is one divided by the number of values in the data set). While this seemed perfectly reasonable to the engineer, he realized that he would have to produce a plot that the boss would accept. To do this, he decided to change his way of thinking about the precision of the data he was presenting.

Rather than considering each value individually, he tumbled to the conclusion that he will have to consider the measurements in "bins" or "buckets" or "groups." Figure 10.3 is the plot which results from considering the data in 6-inch-sized bins.

The boss was happy with the plot of heights so the engineer too was happy, but something still bothered him about the "number of bins" thing. He decided, however, to forget about it for the time being because the boss had given him a new assignment.

The new task is to analyze some financial data. He is to get from the accountants data that show the range of profits the company, from preliminary estimates, might

What is a Distribution?

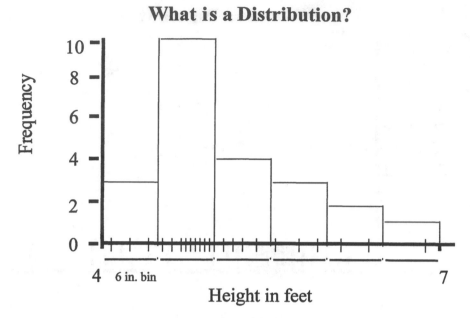

FIGURE 10.3   Plot of "binned" height data.

make from the plant expansion. The question to be answered is, How likely is it that the company will realize, in the first year of production, a profit of around $5MM (the first-year economic hurdle for the project)?

After retrieving the data from the accountants, the engineer loaded the data into statistical software and made three plots. These three plots are shown in Figure 10.4.

When five bins are used to represent the data, the bar chart has a roughly log normal appearance; that is, the plot is skewed to the right (remember, skewness direction is the direction in which the "tail" goes). When 10 bins were used, the resulting plot indicated that there is a wide range of values of about the same frequency with less frequently occurring values on either side. When 40 bars were used, the plot takes on a bimodal appearance.

The changing appearance of the plot, however, is not the most disturbing aspect of the plots. The boss wants to know what is the probability of generating a profit of about $5MM. According to the five-bar plot, the probability of making about $5MM is about 50%. According to the 10-bar plot, the probability is about 30%, and the 40-bar plot indicates that the probability is no greater than 9%.

It is clear from the plots that this assignment must be more clearly defined. Just what did the boss mean by "about $5MM?" If she meant the chance of achieving exactly $5MM, and $5MM just happened to be one of the X-axis values in the data set, then the probability of getting $5MM is 1 divided by the number of values in the data set. If $5MM is not one of the X-axis values plotted, then there is no chance of getting exactly $5MM.

However, the boss said "about" $5MM, so the engineer did not think she meant exactly $5MM. Without clarification concerning the definition of "about," he could

FIGURE 10.4   Frequency plots of common data using 5, 10, and 40 bins.

not determine the level of precision (i.e., bar or bin width) with which he should represent the data. It should be clear from the three bar charts in Figure 10.4 that the probability of achieving "about" $5MM depends on the bin size (bar width) which, in turn, is a function of the definition of "about."

The lesson to be taken from this exercise is that when constructing a contributing factor diagram, the group building the diagram needs to discuss and arrive at a consensus decision regarding the level of precision with which the data need to be considered.

So, to return to the original question posed at the beginning of this section, just what is a distribution? The distribution actually is the scattering of data points along the X axis, as shown in Figure 10.1. A bar chart is a means of presenting the data, but it should be apparent that a presentation of the data should be commensurate with an understanding of the level of precision required.

For example, in our door building example, because the lumber is cut in 1-foot increments (5-feet long, or 6-feet long, etc.), it does not matter to the boss whether the average height measured is 5 feet 5 inches, or 5 feet 7 inches, etc. She makes the simple deduction that she needs lumber at least 6-feet in length. However, if the engineer's charge had been to make space suits for NASA instead of doors, this amount of imprecision in the data could not be tolerated (each suit is built specifically for each astronaut), and tolerances in height are very "tight."

A bar chart, then, is a representation of the distribution. The X-axis of the bar chart represents the range of possible values for the variable being graphed. X-axis values denote the probability associated with a given "bin" size. The bin size should represent the level of precision required. When the contributing-factor diagram is being constructed, the group should discuss and come to consensus concerning the level of precision required and should agree that all plots made for the "answer" variable with respect this consensus. If this decision is not adhered to, very different representations of the data, such as those shown in Figure 10.4, can result. See the Output from Monte Carlo Analysis — The Frequency and Cumulative Frequency Plots section of Chapter 9 for information on converting frequency plots to cumulative frequency space.

DISTRIBUTIONS — HOW TO APPROACH THEM

Those versed in risk assessment and/or statistics tend to classify distributions by their properties. Such names as "normal" (or Gaussian), "Poisson," "gamma," and others are typical. Although this type of jargon is required if you want to be "one of the boys" and proliferate the erudite haze that shrouds statistics, I have seldom found such terms to be useful. Their limited use stems not so much from the fact that these names are relatively meaningless with respect to the actual distribution shape, but rather from the fact that my business has always been to bring risk assessment to people who are not necessarily versed in statistics or risk. Lawyers, accountants, plant managers, engineers, geoscientists, chemists, and others all are very well read with respect to the nuances of their given fields. Many of the people in these disciplines, however, can and do need to perform risk assessments. So while it is true that most of these people are capable in their fields, they likely are not intimately familiar with the fundamentals or particulars of statistics or risk assess-ment. I always liken it to driving a car. I think I can drive pretty well, but I really don't know much (anymore) about what goes on under the hood of my car. I count on someone else to take the complicated mess and make it easy for me to use, because I have to use the car.

That is my job with respect to statistics and risk assessment. Most people need to use these tools but, like the car, they really don't want to know what is going on under the hood, nor do they want to be asked to demonstrate a skill level or knowledge that is outside their ken. I like to say in my risk assessment classes that it is easy to make something hard, but it is really hard to make something easy. So it is with statistics and risk assessment.

I learned early in my career that if I were to be successful in bringing risk assessment to "the masses," then I would have to find a way to insulate and isolate others from the rigors and details of statistics. This required not only an intimate knowledge of the subject manner on my part, but the development of algorithms, processes, and techniques that would allow those not steeped in statistics to use a risk assessment package and get the job done without having to provide "unnatural" (for them) information or to demonstrate extraordinary skills. I also learned early on that it is far better to use your intuition, experience, and knowledge of the subject to build a distribution shape that you think best represents the data or variable. Representing a variable's data by algorithm (i.e., selecting a gamma, or beta, or chi-squared, distribution, etc.) likely is less useful. In most instances, users should be allowed to enter the data that they know about and should be able to shape the distribution to fit their mental model of the actual (parent, if you like) distribution of data. For example, an accountant may know (or think he knows) the distribution of product prices for the next 6 months. Rather than being asked to provide an "alpha1" and "alpha2" value or some combination of mean and standard deviation, variance, kurtosis, or any number of other (to them) meaningless statistical values, the user should be able to input data with which he is familiar (price data, in this case). He/she should be able to shape the distribution to fit historical price data, price projections, or experience-driven intuitive knowledge concerning the distribu-tion shape. The algorithms I developed to build distributions do just that without bothering users with a lot of statistical mumbo-jumbo.

The first step toward "statistics-free" risk assessment was the decision that I should strive to develop algorithms that would allow the generation and presentation of distributions by their general shape instead of by their statistical names. The names usually are tied to their properties (or inventors). To this end, algorithms were designed to allow the user to enter some combination of minimum (min), most likely (ml), maximum (max), and "peakedness" values that would combine to create (mostly) "tailed" distributions. The names of the distributions denote the distribution shape. The primary distributions for which algorithms were developed are

- Symmetrical
- Skewed
- Spike
- Flat

- Truncated
- Discrete
- Bimodal

Reading data (and randomly sampling the data) from a file, peakedness, and, as I like to refer to it in my classes, the "evil, devil's" triangular distribution will also be discussed in this section. I have taken to referring to the symmetrical, skewed, spike, flat, and truncated distribution types as "simple" distributions, not because their mathematics or forms are simple, but because simple methods have been devised for constructing these distributions. The distribution-building algorithms generated can build any of the aforementioned "simple" distribution shapes using select combinations of minimum, most likely, and maximum values. The combination of minimum, most likely, and maximum values needed to construct each distribution type will be conveyed in the following section that describes each type.

SYMMETRICAL DISTRIBUTIONS

The normal (or Gaussian) distribution forms the basis for parametric (classic) statistical techniques such as the "Student's t" test and others. Because this is a risk assessment book and not a treatise on statistics, and in keeping with my bent for practicality, the complex mathematics of such distributions will not be discussed here. In addition, in my quest to prod the reader into considering distribution form rather than function or equation, I will heretofore refer to such distributions mainly as symmetrical.

Some properties of symmetrical distributions (excluding for now triangular distributions which are described elsewhere in this text) include the following:

- They are bell shaped;
- They are symmetrical about the mean value;
- The mean and median values (and the mode, if there are repeating values as in the case of nominal data) are coincident.

In Figure 10.5 is depicted a symmetrical distribution. As previously stated, this distribution, type forms the basis for many classical statistical techniques. However, I have, except in rare instances, found relatively little use in the world of risk assessment for the symmetrical distribution. This is not to pooh-pooh the statistical prowess and significance of the symmetrical distribution — it is the foundation upon which classical parametric statistical techniques are built. Now, having hopefully placated the statistical police with the preceding statements, I will now comment on the use of symmetrical distributions in practical risk assessment.

When building a distribution, the model builder typically is attempting to represent a data set that they have not actually measured. That is to say that the distribution is a representation of what the actual scatter of values would look like had the person actually taken a statistically significant number of measurements for the variable. Having performed literally hundreds of risk assessments, I can assert

FIGURE 10.5   Symmetrical distribution.

with some confidence that application of a symmetrical distribution rarely is called for.

Experience with many individuals and groups has taught me a few things. Many times, a person performing a risk assessment needs to generate a distribution to represent a variable and does not have a credible notion as to the distribution shape that should be used. In such cases, there is a pervasive tendency for that person to "run home to" the symmetrical (normal, if you will) distribution type. I am not quite sure why this is so. Perhaps it is because the Gaussian distribution plays such significant role in classical statistics. Perhaps it is because we typically use the term "normal" to describe such distributions and the term somehow denotes to the person that it is "normal," so it is "normally" what should be used (imagine that!).

In business, science, and other disciplines, those measuring real-world items or events will find that the scatter of actual measurements seldom will be symmetrical. Prices, costs, the size of naturally occurring items (for example, sand-grain size), frequency of occurrence, and other typically measured and ubiquitous risk assessment variables rarely are symmetrical. Even distributions that are used to "expand" or impart "sensitivity" to a single value rarely are symmetrical (i.e., multiplying a single-valued-variable coefficient by a distribution whose minimum, most likely, and maximum values are 0.8, 1.0, and 1.1, respectively, for the purpose of causing the single-valued variable to be able to take on values from 20% less than to 10% greater than the single coefficient). When considering ranges around single values, it seems that phrases like "plus or minus ten percent" just roll easily off the tongues of those who really don't have an idea concerning the range. The values greater than a most likely (or modal) value generally are not distributed exactly as those values less than

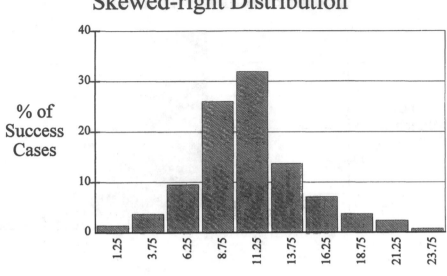

FIGURE 10.6   Skewed-right distribution

the most likely (or modal) value — a symmetry implied and required by the symmetrical distribution.

Figure 10.5 depicts a symmetrical distribution. Symmetrical distributions built by algorithms I have designed are "tailed" (tails of the distributions are asymptotic to the X axis) and are created by supplying a minimum and maximum value only (most likely value left blank). The concentration of values around the most likely value (i.e., the "peakedness" of the distribution) is discussed in a later section of this book.

SKEWED DISTRIBUTIONS

The most prevalent distribution type is the skewed distribution. In my practical vernacular, a skewed distribution is defined as one in which the concentration of values (the "hump," if you will, in a frequency plot) is not central. The log normal distribution is probably the most popular example of the skewed distribution type. A skewed distribution is shown in Figure 10.6.

A collection of values representing most naturally occurring, business, or technical parameters will be skewed when sorted and plotted. This type of distribution can be skewed to the right or to the left (skewness is the direction in which the "tail" goes). Figure 10.7 shows a skewed-left distribution. Skewed distributions built by algorithms I have designed are defined by entering a minimum, most likely (where the "hump" will be in the frequency plot), and maximum value. The tails of these distributions, as represented by a line connecting the tops of the bars in a frequency plot, are asymptotic to the X axis.

Skewed-left Distribution

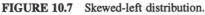

FIGURE 10.7   Skewed-left distribution.

Spike Distribution

FIGURE 10.8   Spiked distribution.

SPIKED DISTRIBUTIONS

In my terminology, a spiked distribution is defined as a "distribution" comprised of a single value. Representation of a parameter by a spike often is handy. Figure 10.8 shows such a "distribution."

Depending upon the question being asked, a single value can be the optimal representation of the answer to that question. For example, if a risk assessment includes a parameter entitled "Number of people tested," this may in fact be a single value. The number of pages that comprise this book is another example in which a single value, rather than a distribution of values, best answers the question.

Spiked distributions also are useful when attempting to determine whether a given risk assessment model is producing the expected results. In a risk assessment, when input parameters are represented by various types of multivalued distributions and are integrated using Monte Carlo analysis (see Chapter 9), the "answer" variables will themselves be distributions. With the resulting distributions being a "smear" of values, it often can be difficult to determine whether the risk assessment model is producing the desired result. In such cases, it often is instructive to represent each of the input variables, for the sake of model testing only, by a spike (single value). So doing will result in single-valued "answer" variables from which it is much easier to determine whether the model is producing the expected results. In the distribution-building algorithms I employ, supplying a most likely value only (leaving the minimum and maximum values blank) will produce a single-valued "distribution."

FLAT DISTRIBUTIONS

What I refer to as the "flat" distribution goes by several other names in the books choked with technical jargon. In such volumes, these distributions are most commonly referred to as "uniform" distributions. In Figure 10.9 is shown a frequency plot that is the result of 1500 "random grabs" from a flat distribution.

Unlike the symmetrical or skewed distributions already discussed, the flat distribution has no concentration of values on the X axis. That is, the X-axis values of

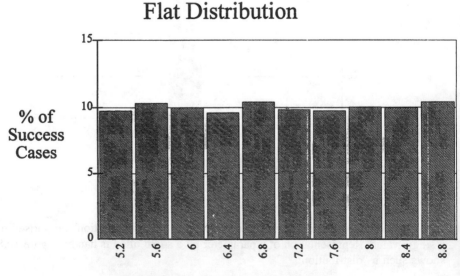

FIGURE 10.9   Flat (uniform) distribution.

such a distribution are "evenly" or uniformly spaced, resulting in a frequency plot (bar chart or histogram) in which the heights of the bars are equal, resulting in a flat (no hump) bar chart. This implies that there is no most-likely value.

The utility of flat distributions is apparent when the user has some idea as to the range of a variable, but has little or no idea what value is most likely. For example, in a given geographic sales area, an area manager may be in the process of opening a new store. For a given product to be sold in the store, the manager knows that in her area the product sells for a minimum of $5.00 per unit (in an economically depressed part of her area) and for a maximum of $9.00 per unit in the "upscale" areas and in airport and hotel shops. The price for the item is set by what the market in the area will bear. The new store is in a newly developed area for which there is no historical data, so the manager may have no idea what the ultimate price of the product will be. She does know, however, that the price will fall somewhere between the just-above-breakeven price of $5.00 and the maximum price of $9.00. In a marketing risk assessment model, the price for this item may best be represented by a flat (i.e., don't know what the most-likely value is) distribution.

The distribution-building algorithms I employ generate a flat distribution when minimum and maximum values are entered (leaving the most likely value blank as with the construction of a symmetrical distribution) and when the user clicks on a toggle button that indicates the distribution is to be flat, rather than symmetrical.

TRUNCATED DISTRIBUTIONS

A truncated distribution is a single-tailed distribution. These distributions can be truncated on the right or left, as shown in Figure 10.10.

Truncated distributions are relevant when the minimum value or the maximum value also is the most likely value. For example, electric motors (furnace motors, pool filter motors, etc.) generally run at their maximum or governed number of revolutions per minute (RPMs). At times, such as when experiencing abnormally high resistance or when "spooling-up or -down" when being turned on or off, the motor may run at less than the maximum RPM. However, running at the maximum RPM is the most likely condition. The RPM situation of such a motor would best be represented by a distribution that is truncated on the right — a distribution in which the maximum value also is the most likely value.

In the oil well drilling business, the restart time for offshore oil platforms (rigs) is a good example of a left-truncated distribution. Oil well drilling and producing activities may be halted on an offshore rig for a host of reasons. The approach of bad weather, a safety hazard, broken equipment, inspections, and a variety of other situations can cause a rig to halt operations. For a given rig in a given area there will be a minimum number of days required to restart operations on the rig regardless of how long or short a time operations were interrupted. Transport of inspectors to the rig, safety procedures, mandated start-up processes, and other required steps combine to prescribe a minimum restart time. For a given rig, the minimum restart time may be 2 days — there is no such thing as a restart time of less than 2 days. This restart time also is the typical restart time. That is, most of the instances of rig shutdown require the minimum number of days (2 in this case) to restart the rig.

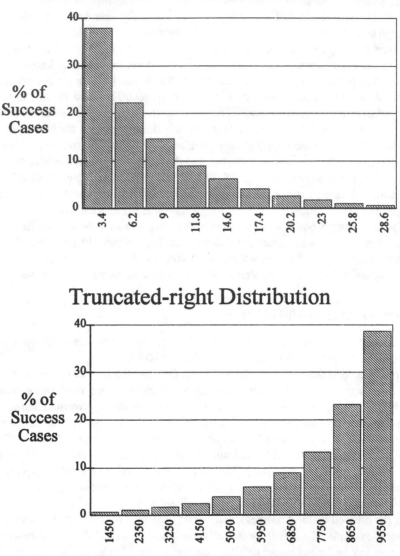

FIGURE 10.10   Distributions truncated on right and left.

However, occasionally operations on the rig are interrupted because of broken equipment that has to be replaced or damage to the rig by inclement weather. Restart times in these and other cases can range up to 30 days or more. Restart times, therefore, are best represented by a left-truncated distribution in which the minimum and most-likely values are coincident (2 days).

Discrete Distribution

FIGURE 10.11   Discrete distribution.

The distribution algorithms I employ require the user to supply a minimum and most likely value (maximum value left blank) for a right-truncated distribution and a most likely and maximum value only (minimum value left blank) to build a distribution truncated on the left. These distributions are "tailed" in that a line connecting the midpoints of the tops of bars in a frequency plot will be asymptotic to the X axis.

DISCRETE DISTRIBUTIONS

The discrete distribution is the first distribution type described here that can't be built simply by supplying some combination of minimum, most likely, and maximum values (and clicking on a "flat" toggle button in the case of the flat distribution). Discrete distributions are of use when attempting to represent a discontinuous variable. For example, the admission price to an event may be $12.00 if you are between the ages of 5 and 65, $9.00 if you are a club member between the ages of 5 and 65, $5.00 if you are a full-time student of any age, and $0.00 if you are over 65 years of age or are 5 years old or under. A risk assessment model may incorporate as an input variable the number of people in each price category who attended the event. Such a distribution is shown in Figure 10.11.

Admission price is a discrete, and not a continuous, distribution. In this distribution there are four discrete values: 0, 5, 9, and 12. When sampling this distribution in Monte Carlo analysis (see Chapter 9), there is no chance of drawing a value of, say, 6 from this set of values. Many real-world parameters used in risk assessment models are best represented by a discrete distribution.

The distribution-building algorithms I use will generate a discrete distribution when the user enters a distribution value (like the 0, 5, 9, and 12 in our example)

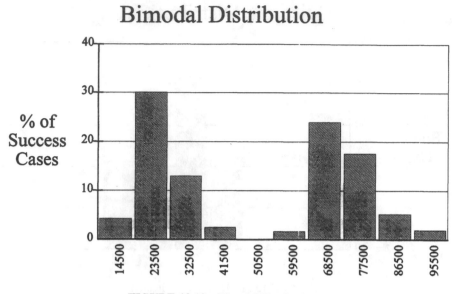

FIGURE 10.12   Bimodal distribution.

and a weight for each value. The weights entered are used to determine the relative frequency of the discrete values.

BIMODAL DISTRIBUTIONS

A bimodal distribution simply is one that contains more than one most likely value (i.e., more than one "hump" in a frequency plot, or, in statistical jargon, more than one modal value). Such a distribution is shown in Figure 10.12.

The bimodal distribution depicted in Figure 10.12 is typical in that it appears to be essentially two skewed distributions plotted in the same X-Y space. Such a distribution may be invoked to represent, for example, the return on an investment that depends upon the participation of other key investors. If we are to carry the investment alone, then the return may be in the lower range with a most-likely return of around $23,500. However, if we are successful in attracting other investors, our return may be in the higher range with a most likely return of about $68,500.

As previously indicated, a bimodal distribution is essentially two distributions plotted in a common X-Y space. The two distributions need not be skewed as shown in Figure 10.12. Nearly any combination of distribution shapes can be used to create a bimodal distribution. For example, we may have an investment that has the potential to return between $2 million and $5 million if the financial deal does not "fall through." If the deal falls through, no return will be realized. There may be a 20% chance that the deal may fall apart, thus, a 20% chance of $0 return. In the bimodal distribution shown in Figure 10.13, there is a 20% chance of drawing (in Monte Carlo sampling — see Chapter 9) a 0 value and an 80% chance of drawing a value of between $2 million and $5 million. This distribution is a combination of spike and skewed distributions.

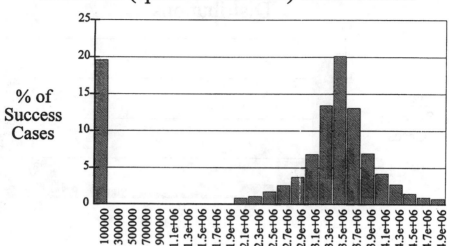

FIGURE 10.13 Bimodal distribution comprised of spike and skewed distributions with associated weights of 0.2 and 0.8, respectively.

In yet another scenario, we may be attempting to represent with a bimodal distribution the amount of water processed by a water treatment plant on a given date. The plant may be in the process of being enlarged with the addition of new intake and output pipelines. If the construction is not finished on the date in question, then the plant will run at its old capacity. If, however, the construction of the new pipelines is complete or partially complete on the date in question, then the plant could process water at a rate somewhere between the old processing rate and the new one. However, the most likely scenario is that the plant will process water either at the old rate or at the new rate. In Figure 10.14, two truncated distributions (one truncated on the left, the other truncated on the right) were used to create a bimodal plot (note that this plot is not unlike a "beta" distribution).

Most bimodal distribution-building technologies offered by risk assessment packages offer a means of "emphasizing" one of the distributions (modes) relative to the other. In the algorithms I use, a "weight" can be assigned to each of the two distributions that compose the bimodal. The weight values simply control the proportion of X-axis values assigned to each of the two distributions that are combined to produce the bimodal distribution. In the case of the distribution shown in Figure 10.13, for example, a weight of 0.2 (20%) was assigned to the 0 (spike) value and a weight of 0.8 (80%) was assigned to the skewed distribution.

READING DATA FROM A FILE

Distributions are used to represent variables because rarely have we measured the entire parent population (i.e., we rarely have made all the measurements it is possible to make). Also, distributions are a convenient method of representing an entire

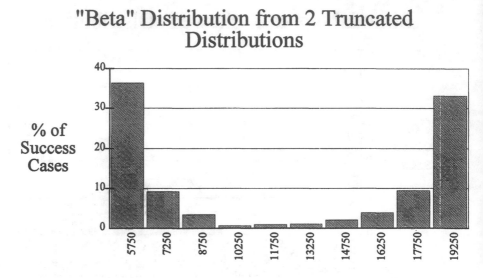

FIGURE 10.14   Simulation of a beta distribution by combining two truncated distributions in a bimodal distribution.

population of values by requiring just a few input values from the person supplying the data (minimum, most likely, and maximum, or mean and standard deviation, or some such set of values).

At times, however, complete or nearly complete sets of data exist that may adequately represent the variable. In such cases, it may be tempting to feed this data directly into the risk model so that the Monte Carlo random-sampling engine can use the actual data as the basis of its sampling. There is nothing technically foul about sampling from "raw" data rather than from distributions, but there are a few potential problems that might arise.

Distributions built by software systems generally are composed of a discrete number of values. That is, between the minimum and maximum values of a distribution, there typically are 100 to 500 discrete values (this changes with both software package and distribution type) from which Monte Carlo analysis can randomly select. These values usually do not exist as a physical array of values, but are algorithmically generated at run time. Nonetheless, there typically are a limited number of values from which to sample.

One of the factors that dictates the number of Monte Carlo iterations required in a stochastic risk assessment model is the number of values that comprise the distributions. The greater the number of values, the greater the number of iterations required to adequately sample the distributions (techniques such as Latin Hypercube can reduce the required number of iterations, but these techniques have problems of their own). The data values read from a file will be used as the basis for Monte Carlo sampling — just like a distribution. When files containing large numbers of discrete values are employed, it should be obvious that a concomitantly large number of Monte Carlo iterations will be required to adequately sample the array of values in the file. This can have a significant impact on run time and risk-program complexity.

Having just warned about the foibles associated with files containing large numbers of values, I now will address some of the problems associated with files that contain a relative dearth of values. It is a relatively rare occurrence that data collected over a finite length of time adequately represent reality. For example, an engineer on an offshore oil drilling platform may have recorded in a file the number of days the platform is shut down ("downtime") each time there is a problem of great enough significance to halt operations. The engineer may have been recording data for 5 years. Typical downtimes have been in the range of 2 to 6 days per event. However, in the third year of downtime monitoring, a large "100-year" wave (a wave of this size is statistically expected to occur only once in 100 years) struck the platform, destroying critical equipment. The resulting downtime was 30 days. This value of 30 is one of only ten values that have been recorded over the 5-year period. Monte Carlo random selection will, on average, select the value of 30 one time out of ten. This, effectively, transforms the 100-year wave into the 10-year wave.

Except in the case of "short time" events, it is a rare situation that we have collected data over a sufficient length of time for our amassed data to represent the real-world situation. Great caution should be exercised when substituting data files for distributions. Make every effort to ensure that a sufficient number of values has been collected so that the range and concentration of the values adequately aligns with the real world.

PEAKEDNESS

From the never-ending battle to keep things simple and practical, the notion of peakedness of a distribution was born. The typical user of a risk model is not versed in the concepts and nuances of statistical analysis including the concept of kurtosis. Kurtosis can be loosely (but not technically) defined as a measure of the "peakedness" of a distribution which is, in turn, a reflection of the "concentration" (density distribution) of values on the X axis of a frequency plot. Terms such as "platykurtic," "leptokurtic," and others are used to describe distribution shapes. The kurtosis of a Gaussian (normal) distribution is 3.

Risk model users usually do not want to be bothered with this. What they would like to do is simply shape the distribution the way they wish without having to supply mean, standard deviation, kurtosis, or other statistical data. The concept of peakedness is a means by which they can do this.

The peakedness algorithm I employ (which can be replicated in any home-grown risk assessment package) is nothing more than an algorithm that controls the concentration of X-axis values around the most likely value (the "hump" in the distribution) in those distributions that have most likely values. We explain peakedness to the user as being their confidence in the most likely value.

A peakedness range of 0 to 10 has been established. A slider bar is utilized to select a value between 0 and 10. A peakedness of 10 concentrates values near the most likely value, resulting in a frequency plot like that shown in Figure 10.15. In this distribution, the probability of Monte Carlo random sampling selecting a value at or near the most likely value is relatively great compared to the probability of selecting a value in an equally sized range in the tail area of the distribution. By

Distribution with Peakedness of 10

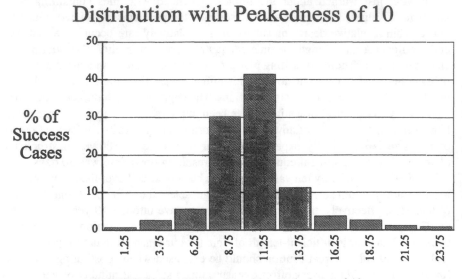

FIGURE 10.15   Distribution with a peakedness of 10.

Distribution with Peakedness of 0

FIGURE 10.16   Distribution with a peakedness of 0.

selecting a peakedness value of 10, the user is expressing relatively high confidence that values near the most likely value are those very likely to represent the variable.

A peakedness of 0 puts more values in the tails of the distribution at the expense of the most likely value. Figure 10.16 shows such a distribution. In a 0-peakedness distribution, values near the most likely are still more likely to be selected, but values in the tails of the distribution are relatively likely to be selected compared to a

FIGURE 10.17   "Tailed" distribution contrasted with a triangular distribution.

distribution with a peakedness value of greater magnitude. A low peakedness value indicates that the user has relatively low confidence in the most likely value. Using this peakedness scale, the user can express their confidence in their data and in so doing, control a major aspect of distribution shape.

TRIANGULAR DISTRIBUTIONS

Triangular distributions are ubiquitous in risk assessment models not because they best represent the data, but because they are, in most software packages, easy to build. The typical distribution-building and Monte Carlo-based risk assessment software package requires the user to enter things such as means and standard deviations (and other values that will be covered in the "Specific" Distribution Types section of this chapter) to build distributions. Most users are, at least, reluctant to supply such data and so turn to the triangular distribution that requires only a minimum, most likely, and maximum value for its construction. This is very attractive.

The fundamental problem with the triangular distribution, however, is that nothing typically incorporated in a risk model is distributed in a triangular fashion. This is more than a cosmetic problem.

Figure 10.17 contrasts a triangular distribution with a "tailed" (asymptotically) distribution. It can be seen in the plot that the probability of occurrence of values in the tails of the distributions (as well as the most-likely value) are significantly different. In the case of the plot in Figure 10.17, in which X-axis values represent heights of adults, the triangular distribution suggests that about one out of two adults is near 7 feet in height (frequency of that height relative to the most-likely height). We know that not every second person we meet on the street is of this great height.

FIGURE 10.18   Beta distribution.

By contrast, the asymptotically-tailed distributions suggests that people of great height are relatively rare.

This discrepancy may seem academic, but the probability exaggeration in the tails of triangular distributions can lead to significant error, especially in the area of portfolio analysis. If the X-axis values in Figure 10.13 were dollars returned on a single investment (instead of height), the combination of multiple distributions in a portfolio-management scheme would lead to significantly inflated "upside" returns. For this reason and others, triangular distributions should be avoided unless they fit the real-world situation better than any other distribution type.

"SPECIFIC" DISTRIBUTION TYPES

For lack of a better term, I label as "specific distributions" all distributions that have been assigned names derived from their authors, their function, the algorithm that generated them, and other sources. Some of the most popular distributions of this category are the chi-squared, beta, gamma, Weibull, exponential, error function, and others.

A common thread of kinship among these distributions is that they require the user to supply distribution-building values that, typically, are difficult for the average person to provide or understand. The beta distribution (effectively two truncated distributions) in Figure 10.18 is an example. Construction of a beta distribution usually requires the user to provide alpha 1 and alpha 2 parameters (well, isn't that nice!). A gamma distribution will require you to provide alpha and beta values, a chi-squared distribution needs a nu value, and so it goes.

These distribution types can come in handy. For example, engineers may know from generations of experience that a given parameter is best represented by a chi-squared distribution. If this is so, then a chi-squared distribution should be used. However, in most situations, the average user will be at a loss to supply the statistical information required to build such distributions, and a "simple" distribution (as defined previously in this chapter) is a better choice.

11 Chance of Failure

CONTENTS

Chance of Failure — What Is It?..129
Failure of a Risk Component..130
Chance of Failure That Does Not Affect an Input Distribution.........................133
Incorporating Chance of Failure in a Plot of Cumulative Frequency.................134
Another Reason for Chance of Failure..136
The "Inserting 0s Work Around"...138
Chance of Failure and Multiple Output Variables...139

CHANCE OF FAILURE — WHAT IS IT?

I will begin this chapter by stating that there is no more common cause for spurious risk assessments and misinterpretation of risk-model results than the exclusion of chance of failure from the analysis. It will be illustrated in this chapter that consideration and integration of chance of failure, when appropriate, is crucial.

Chance of failure (COF), as I write about and define it here, is the probability that the object of the risk assessment will experience abject failure. The total chance of success (TCOS) for the object of a risk assessment is calculated thus:

TCOS = (1 − COF variable 1) × (1 − COF variable 2) × ...(X1 − COF variable N)

where N is the number of variables that have an associated COF. Given this definition and calculation of TCOS, it should be clear that the application of the COF concept is valid only in cases in which failure of any variable that has an associated COF will cause the object of the risk assessment to fail. A physical analogy is the proverbial chain made of links that may be thought of, for example, as our project. A project risk assessment is composed of variables. If one link breaks, the chain fails. In the COF scheme, if one variable fails, the project fails. So, the concept of COF, as presented here, is *not* valid in situations in which failure of one component does not necessarily cause the entire enterprise to fail. A physical analogy is the strands of a rope. Failure of one strand (unlike failure of one link in the chain) does not necessarily cause the rope to fail.

A good real-world example in which COF might be applied is the launching of a new product in the marketplace. We may be considering, among other things, the price we need to get, legislation we need to have passed for our product, and the labor costs. If any one of these variables turns out, in the real world, to be outside our acceptable limits, then the product introduction will fail. That is, if we need a price of at least $12.00 per unit but all the market will bear is $10.00, then the game is over. Likewise, if the government fails to pass the required legislation, we fail. Similarly, we may experience labor costs above what we can bear, causing the enterprise to fail (no matter what other values we experience for other variables).

In contrast, some situations do not warrant the consideration of COF. For example, for our corporation to survive, we may need to attain an after-tax profit of $10MM. We arrive at our corporate profit by combining the profits from the 17 companies that comprise the corporation. If, in a given year, one or more of the companies "fails" (i.e., does not realize a profit), this does not necessarily spell failure for the corporation as long as the required profit can be gleaned from the companies that did show a profit.

A good real-world hybrid example is your car. Almost all components of your car have an associated COF. However, if we are building a risk assessment model to calculate our chance of the old car making a 1000-mile trip, consideration of failure for certain components is critical (transmission, alternator, radiator, etc.) while failure of other components (radio, electric locks, etc.) may simply be annoying.

FAILURE OF A RISK COMPONENT

Well, now that I have expounded a great deal about failure of a component or variable in a risk assessment, just how is it that failure is defined for a variable? The risk-model builder has to generate a definition for failure for each variable that has an associated COF. Let us begin with the situation in which COF affects the distribution.

Consider the simple case of the environmental engineer charged with supplying potable water in a drought-stricken area. In Figure 11.1 are shown two distributions: one for sediment load (percentage of a liter of water comprised of suspended solids) and one for size of the underground water reservoir (measured in cubic meters). The percentage of rock particles in a liter of water in the given area has, according to historical data, ranged from a minimum of 3% to a maximum of 20%, with a most-likely sediment load of 10%. However, experience has shown that sediment loads greater than 14% are too high, causing the pipes to clog.

The percentage of the histogram bars (frequency) in Figure 11.1 that will fall above 14% is the COF for sediment load. COF generally is expressed as a percentage and is either calculated by the computer algorithm (as in the risk assessment package I use) or estimated by the user. The blackened area in the percent-load plot represents 10.5%, or a COF of 10.5. In this case, failure of the sediment-load variable is defined as the probability that the actual sediment load value will be greater than our maximum acceptable value of 14%.

Now let's turn to reservoir size. The size of reservoirs in the area have ranged from a minimum of 3 million cubic meters (MCM) to a maximum of 90 MCM, with a most likely size of about 35 MCM. Economic calculations indicate that if the reservoir is less than about 20 MCM, then the reservoirs contain too little water to be economically developed. In Figure 11.1, the blackened area in the reservoir size histogram represents 5.5% of the area under the curve and, thus, a COF of 5.5.

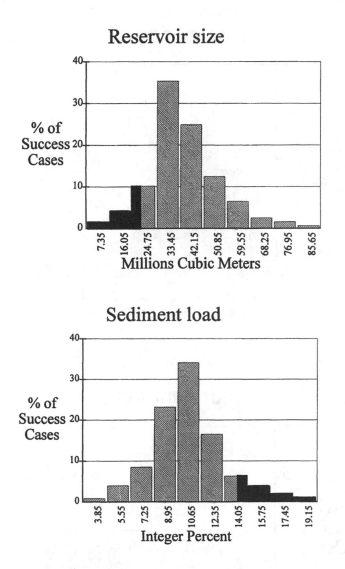

FIGURE 11.1 Frequency plots for sediment load and reservoir size. Black areas represent % chance of failure (above 14% for sediment load and below 20 MCM for reservoir size).

Failure for the reservoir size variable is defined to be the probability that the actual reservoir size will be less than our acceptable minimum size of 20 MCM.

In a risk assessment incorporating sediment load and reservoir size, the respective input distributions for the two variables would appear as those shown in Figure 11.2. Both of these distributions represent "success-only" values. Note that it is important to not allow the probability of the maximum-allowable sediment load or the minimum-acceptable reservoir size to go to near-zero probability.

The original sediment-load and reservoir-size distributions have essentially been "clipped" to produce the distributions shown in Figure 11.2. The parts of each

FIGURE 11.2   Sediment load and reservoir size distributions that have been "clipped" by their respective chances of failure.

distribution that have been clipped will be used in the calculation of the TCOS of the project thus:

$$TCOS = (1 - 0.15) \times (1 - 0.055) \qquad (11.2)$$
$$= .85 \times 0.945$$
$$= 0.803$$

CHANCE OF FAILURE THAT DOES NOT AFFECT AN INPUT DISTRIBUTION

In our water project example, one of the input variables is the size of the reservoir. Thus far, this variable has a single associated COF defined to be the chance that the reservoir is of a size that would yield uneconomical volumes of water. However, in the area where our water well is to be drilled, there is some chance that the desired reservoir rock type may have been altered by subsurface hydrothermal fluids (essentially, underground flowing hot water). The fluids may have transformed the reservoir to a rock type that is not of reservoir quality and, therefore, can yield no water at all.

It may be that we have measured the size of the potential reservoir rock just fine (using "remote sensing" techniques such as seismic or gravity-related technologies), but there is a chance that we are measuring the size of a "reservoir" of the wrong rock type. Therefore, we may define a second COF associated with reservoir size. This additional COF might be defined as the chance that we are measuring the size of "the wrong thing" (i.e., the "reservoir" is composed of a rock type from which we can't extract water at all). This COF is associated with the reservoir size distribution, but does not clip or in any other way affect the distribution. For the sake of this example, let's say that we believe that there is a 10% chance that we have accurately measured the size of a body of rock of the wrong type.

Yet another variety of COF that does not influence input variable distributions is the type that is not associated with any input distribution at all. Sticking with our water project example, there may exist yet another reason for potential failure of the project. For our project to succeed, we need the local legislature to pass a law that exempts our project from regulations which preclude drilling in our proposed project area. Even though this project is being pursued at the behest of the local population, there is a significant chance that environmental groups may successfully block passage of the required legislation.

In this situation, we may define for our water project risk assessment model an input variable that is a COF only (no associated distribution). The COF-only variable, in this case, might be defined as the chance that we will fail to get the required exemption. Because this variable has no distribution, it will not affect the Monte Carlo-calculated X-axis values but will affect the "elevation" of X-axis values to form our cumulative frequency plot and the projected X-axis intercept of the curve. Let's assume that we believe there is a 20% chance that the required legislation will fail to be enacted. Given the two additional COF values we have added to our project, the TCOS for the project would now be as follows:

$$\text{TCOS} = (1 - \text{COF sediment load}) \times (1 - \text{COF reservoir size}) \quad (11.3)$$
$$\times (1 - \text{COF wrong rock type}) (1 - \text{COF legislation})$$

$$= (1 - 0.15) \times (1 - 0.055) \times (1 - 0.1) \times (1 - 0.2)$$

$$= 0.85 \times 0.945 \times 0.9 \times 0.8 = 0.578$$

INCORPORATING CHANCE OF FAILURE IN A PLOT OF CUMULATIVE FREQUENCY

In the preceding example, our total chance of success for the project is 57.8%. As described in the Output from Monte Carlo Analysis section of Chapter 9 of this book, the usual process for creating a cumulative frequency plot begins with sorting the X-axis values from smallest to largest. The Y axis is subdivided into a number of equally spaced segments (the number of segments equal to the number of Monte Carlo iterations) and the X-axis value of greatest magnitude is "elevated" one probability unit, the next largest value is "elevated" two probability units, and so on until the leftmost X-axis value is "elevated" N (number of iterations) probability units (Figure 11.3).

In the immediately preceding section of this chapter, we calculated the TCOS of the water project to be 57.8%. To incorporate COF in the cumulative frequency plot, rather than subdividing the 0–100 Y-axis range into N (number of iterations) segments, the 0-TCOS (in this case, 0–57.8) range is subdivided into N equally spaced segments. The X-axis values are "elevated" as described previously, resulting in a cumulative frequency plot whose projected "left" end would intersect the Y axis at the TCOS value (Figure 11.4).

FIGURE 11.3 Individual X-axis values "raised" parallel to the Y-axis according to their cumulative probability.

Dollar Benefit from Water Project

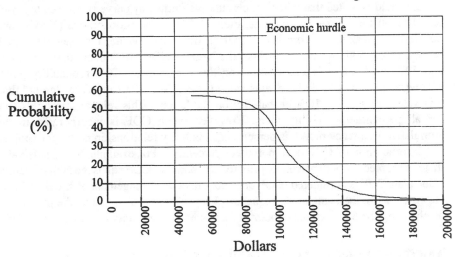

FIGURE 11.4 Cumulative frequency plot that includes chance of failure.

Dollar Benefit from Water Project

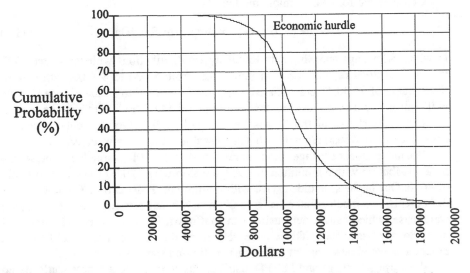

FIGURE 11.5 Cumulative frequency plot equivalent to that in Figure 11.4 but without chance of failure.

Figure 11.5 shows the same water project data plotted without considering COF. Note that the probability of achieving a dollar benefit amount of $90,000 or greater is about 86%. When considering COF, as shown in Figure 11.4, the probability of achieving the same dollar benefit is only about 50%. This is a significant difference that should result in different decisions being made concerning the water project.

It should be noted that when the cumulative frequency curve is allowed to "rise" to 100% on the Y axis (no consideration of COF), the probabilities (Y-axis) and percentiles (X-axis) are correlated. That is, the X-axis value, for example, which represents the 50th percentile (50% of the X-axis values are "larger" than this value and 50% are "smaller") also projects on the Y axis at the 50% probability point. Likewise, the 90th percentile X-axis value (i.e., 10% of the values are "larger" than this value) projects at 10% probability on the Y axis. This relationship holds true for all percentile/probability pairs. However, when COF is incorporated in the cumulative frequency plot, this percentile/probability relationship no longer applies.

Incorporation of COF does not affect percentiles. For example, the X-axis value that represented the 50th percentile prior to incorporating COF still is the 50th percentile value after COF is considered (COF does not affect the arrangement of X-axis values). However, when the cumulative frequency plot is "depressed" by COF, the probability of obtaining the 50th percentile point (or a value "greater" than it) no longer is 50%.

ANOTHER REASON FOR CHANCE OF FAILURE

In the water project example, the cumulative frequency curve was appropriately depressed to account for COF. This is all well and good; however, let's consider the consequence of ignoring COF in a simple risk model in which COF does influence the input variable distributions.

Consider the following simple model:

$$\text{Income} = \$/\text{Package} \times \text{Shipment Rate/Day} \qquad (11.4)$$

in which \$/Package and Shipment Rate/Day are distributions with associated COF values. The \$/Package distribution may range from \$1.00 to \$10.00 with a most likely value of \$6.00. The associated COF will be defined to be the chance that we will realize a \$/Package rate less than the required \$4.00. Below this rate, a partner company that shares in the \$/Package rate will pull out of the enterprise and, consequently, our business will fail. This distribution appears in Figure 11.6.

The independent package shipment rate per day (PSRPD) variable is represented by a distribution with a minimum of 2000, a maximum of 20,000, and a most likely of 8000. Our business could survive with a minimum PSRPD of 2000; however, if the PSRPD gets much greater than 14,000, a significant bottleneck is created at the warehouse, which causes our business to shut down for an unacceptable length of time. So, the associated COF will be defined as the chance that the PSRPD will exceed our warehouse capacity. This distribution appears in Figure 11.7.

If we use \$/Package and PSRPD distributions as inputs to a Monte-Carlo-based risk model that calculates income from the (admittedly oversimplified) equation

$$\text{Income} = \$/\text{Package} \times \text{PSRPD} \qquad (11.5)$$

then we could, on these Monte Carlo iterations, draw the following \$/Package and PSRPD values:

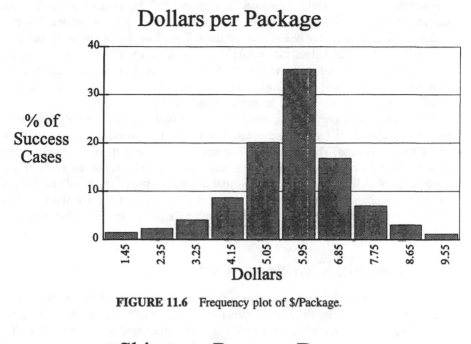

FIGURE 11.6   Frequency plot of $/Package.

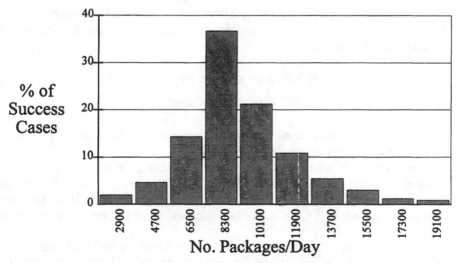

FIGURE 11.7   Frequency plot of package shipment rate per day.

$$\text{\$/Package} \times \text{PSRPD} = \text{Income} \tag{11.6}$$

$/Package	PSRPD	Income
3	10,000	$30,000
5	6,000	$30,000
2	15,000	$30,000

Each of the random selections draws $/Package and PSRPD values that result in an income of $30,000. However, two of the three iterations include values that fall within the failure range for one of the variables. Therefore, given our three iteration case, two thirds of the values that would plot at $30,000 on a histogram of income would not have succeeded in the real world — the histogram would significantly exaggerate our chance of achieving $30,000 (in this instance, our chance of achieving $30,000 is only one third of the histogram's indicated probability).

It is tempting to suggest as a solution to this problem that we simply "clip off" the failure parts of the input distributions, but that is tantamount to hiding our heads in the statistical sand. The values that populate the failure sections of distributions such as these do occur in the real world and have got to be reckoned with. When the concept of COF is employed, the distributions are "clipped" to exclude the failure values (sometimes the failure area can occur in the middle of the distribution), but are accounted for by "depressing" the cumulative frequency curve so that probabilities interpreted from the curve reflect the COF.

THE "INSERTING 0s WORK-AROUND"

To nearly everyone who comes to grips with the concept of COF (especially spreadsheet bigots, it seems) it eventually occurs that, "Hey, I could just calculate a 0 when I run into a "failure" situation and I won't have to fool around with this COF stuff." I can tell you from experience that the route of "substituting 0s for failure" is the road to ruin.

For example, it is tempting to consider inserting some logic in the risk model that will recognize a failure value and so calculate a zero. In our $/Package and PSRPD example you might write

$$\text{If } \$/\text{Package} < 4 \text{ then Income} = 0 \qquad (11.7)$$

$$\text{If PSRPD} > 14 \text{ then Income} = 0$$

and think you've taken care of that nasty failure problem. Well, there are several things wrong with this approach.

First, some risk assessment models are quite complex and can consume a significant amount of time and resources for each iteration. Typically, it takes no less time to generate a zero result than to generate a non-zero result. If the situation includes a relatively high COF, most of the time and effort is spent generating noncontributing 0s.

In this same vein, if a 1000-iteration risk model includes a 50% COF, this means that 500 0s will be generated and that the cumulative frequency curve will be composed of only 500 "nonfailure" values. This wouldn't be so terrible if each situation you needed to run through the model had exactly the same COF. It is the usual situation, however, that individual projects have unique associated COFs. Using the zero-insertion method, you will be left with a curve comprised of 500 points (50% COF) from one run of the model, a second curve made up of 300 points (70% COF) from a second run of the model, and so on. Not only are the curves composed of relatively few values, but the disparity in values can make statistical comparison of curves difficult or invalid.

The clever person might look at this disparity of nonfailure values problem and conclude, "Hey, I know what I'll do, I'll just let the model run until it generates, say, 1000 non-zero values." Nice thought. More iterations generate more zeros. More iterations also increase the chance of sampling and combining extreme values in the "tails" of the distributions. Under this "more iterations" scheme, two output cumulative frequency curves generated by combining the same input distributions but having different TCOS values can have significantly different X-axis ranges. This is because more iterations tend to extend the end points of the cumulative frequency curve. This can make it difficult to compare curves from multiple runs of the risk model. A cumulative frequency curve generated using the COF concept will always contain the same number of points as the number of Monte Carlo iterations.

As is my wont, I have saved the best for last. The most glaring problem associated with the "substitute 0s" problem is that sometimes zero is a nice benign value that can be used as a failure indicator and sometimes it isn't.

Consider the typical risk model. There usually exists "up front" a technical section followed by, at least, a section that calculates economics (sometimes an "engineering" section might be sandwiched between the technical and economic parts). In any event, zero might be able to represent failure in one part of a model but would not be interpreted as failure in other sections.

For example, it might be fine to calculate factory output in the first part of a risk model. Factory output is either something (a number greater than zero) or nothing. There likely is no such thing as negative factory output. So, it might be tempting in this case to represent with zero the failure of the factory to produce. An output curve of production, then, might include many zeros indicating the number of times the Monte Carlo process calculated no production.

In a multistep risk model, however, the output from one step is passed along to the next. It may well be that zero can represent failure in the "calculate production" step, but a subsequent engineering step (in which there can be "successful" negative and positive values) or economics step (in which deficit spending does not represent failure) may not treat zeros as the benign values they were in the "production" stage of the model. Passing along strings of zeros from one part of a risk model to another is a dangerous practice.

CHANCE OF FAILURE AND MULTIPLE OUTPUT VARIABLES

A cautionary note concerning the application of the COF concept with multiple output variables. Consider the case in which we are computing two output variables A and B from input variables C, D, E, and F. The equations might be

$$A = C + D \qquad\qquad (11.8)$$

$$B = E \times F$$

Each of the four input variables may have an associated COF. In this case the risk model should calculate two TCOS values; one for A from the COF values associated with C and D, and one for B from the COF values for E and F. Given the separate and independent equations for A and B, the failure of, for example, E has no affect on A. Likewise, the failure of either C or D will not be considered in calculating the TCOS for B. Great caution should be taken when constructing risk models so that TCOS values for individual output variables are correctly calculated.

12 Time-Series Analysis and Dependence

CONTENTS

Introduction to Time-Series Analysis and Dependence...141
Time-Series Analysis — Why?...141
Time-Series Analysis — How?...142
Time-Series Analysis — Results...143
Time-Series Analysis That's Not...144
Some Things to Consider..144
Dependence — What Is It?..147
Independent and Dependent Variables 148
Degree of Dependence 148
Multiple Dependencies and Circular Dependence..152
Effect of Dependence on Monte Carlo Output...153
Dependence — It's Ubiquitous 153

INTRODUCTION TO TIME-SERIES ANALYSIS AND DEPENDENCE

Chance of failure is a primary risk assessment concept and thus is treated in a separate chapter. There are, however, two (at least) other data-manipulation techniques that are critical elements of real-world risk models. Time-series analysis and dependence are functions that help risk models better emulate actual situations.

Time-series analysis allows us to break free of the single-period assessment and to project the analysis through time. This is a facility that is critical in most business applications, for example, in which cash flows over time are combined to generate project-representative values such as net present value (NPV), etc.

Dependence is a technique that allows a risk model to respect the relationships between variables. Without application of dependence technology, any hope of emulating real-world processes is folly. Both time-series analysis and dependence are treated in the following sections.

TIME-SERIES ANALYSIS — WHY?

The desired outcome of a risk analysis is, of course, a stochastic assessment of the situation at hand. In a business setting, regardless of the technical, environmental, political, or other nature of the risk model, ultimate decisions primarily are based on a probabilistic financial assessment of the opportunity. Financial measures such

Year	1998	1999	2000	2001	2002
Labor costs (in thousands $)	900	600	550	500	500
Capital costs (in thousands $)	1000	500	400	400	400
Yearly production (tons)	30000	35000	40000	45000	45000

FIGURE 12.1   Spreadsheet with time periods as columns and variables as rows.

as discounted return on investment (DROI), internal rate of return (IRR), or NPV are typical measures upon which decisions are based. Financial considerations of this ilk require the evaluation of an opportunity over time, commonly some number of years. An NPV, for example, is based upon a series of cash flows.

It is beyond the scope of this book to explain the nuances of the calculation of measures such as NPV, the definition of which can be found in nearly any basic financial text (see selected readings for this chapter). It is, however, precisely the premise of this chapter to explain how a risk model can result in the calculation of yearly cash flows and financial measures such as NPV.

TIME-SERIES ANALYISIS — HOW?

Economists working for most companies these days typically employ spreadsheets of one type or another to handle the financial attributes of a project. It is quite common to see in such spreadsheets some number of columns designated as years with rows below the yearly column headings representing variables in an analysis. Variables such as yearly labor costs, capital costs, yearly production, and other parameters are typical (see Figure 12.1). In a deterministic spreadsheet, these parameters and others would be combined to calculate, for example, a single NPV. In the risk assessment world, however, we strive for a distribution of answer-variable coefficients.

Generating a distribution for an answer variable such as NPV is accomplished by combining input variables that are themselves distributions. Using spreadsheets and a spreadsheet add-on risk assessment package, the yearly and single-valued parameters (i.e., labor costs, etc.) can have each of their yearly values expanded into a distribution. That is, rather than having the 1998 labor costs in Figure 12.1 represented by the single value of 900, 1998 labor costs might range from a minimum of 700, a most likely of 900, and a maximum of 1000.

Several techniques may be employed to transform single values into distributions. For example, it may be specifically known that 1998 labor costs will range from 700 to 1000 and that 1999 labor costs will range from 500 to 750 and so on for subsequent years. In the risk assessment software, then, a separate distribution can be established for each year for each variable. This technique has the advantage of allowing the risk model user to specifically control the parameter values through time. Several disadvantages are inherent in this methodology. First, it is rare, indeed, that we in fact know specific ranges for years beyond the current year. Also, it should be clear that implementing this type of single value expansion puts the onus on the user to build an inordinate number of distributions (one per year per variable). Generally, this is impractical.

Another method commonly employed to expand single values into distributions is to establish one or more "expansion" distributions. For example, we may believe that our labor costs in the first 3 years of the project will range from 10% less to 20% greater than the single value in the spreadsheet. We can then, for the first 3 years, multiply the single spreadsheet value by a distribution that ranges from 0.9 to 1.2. If we believe that for some number of subsequent years different ranges are valid, other expansion distributions can be established and applied in the appropriate time frames. This method of single value expansion reduces the year-to-year control of the parameter distribution, but also reduces the number of distributions the user is required to supply. On each iteration of the Monte Carlo process, the 0.9-to-1.2 distribution will be sampled and the randomly selected value will be multiplied by the single value in the spreadsheet.

Thus far, the time-series examples in this chapter have involved the combination of a risk modeling package and a spreadsheet. However, any complete and self-respecting risk assessment software package should have the self-contained wherewithal to effect time-series analysis without need of a spreadsheet. Such risk packages must contain a means by which any number of time series can be established and a syntax for manipulating multivariate analyses over time. See Chapter 14 of this book for an illustration of such an analysis.

TIME-SERIES ANALYSIS — RESULTS

A risk analysis equation generally terminates in the calculation of a resultant variable that is represented by a distribution. Because a time-series analysis generally considers a proposition over a number of years (i.e., a distribution of years), a resulting time-series answer-variable plot must be able to depict a distribution of distributions. Such a plot is shown in Figure 12.2.

In Figure 12.2, each vertical bar represents the distribution of values for 1 year in a time series. The short horizontal bars at the bottom and top of each vertical bar represent the minimum and maximum values, respectively, for that time period. The third horizontal bar represents either the mean or median value of the distribution. Positioning the computer's cursor on any vertical bar and "clicking" on it can result in the display of the distribution for that time period (Figure 12.3). In this example, a time series of cash flows was generated. From such a time series, an NPV distribution for the project can be calculated.

In reality, a time series plot such as that in Figure 12.2 "knows" nothing about time. X-axis increments can represent any set of time periods such as months, days, hours, etc. In fact, the X-axis intervals need not represent time at all. For example, each tick mark on the X axis might represent a different investment opportunity with

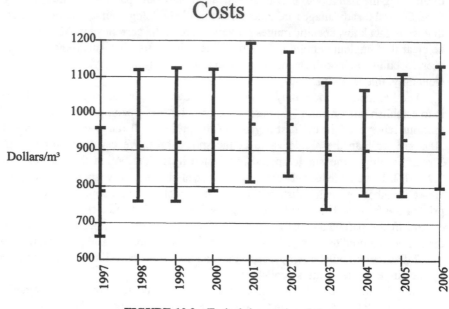

FIGURE 12.2   Typical time-series plot.

each vertical bar indicating the range of return for that investment. An example is given in the next section.

TIME-SERIES ANALYSIS THAT'S NOT

A conventional time-series plot is just that — a plot of a series of time periods. The bars on the plot represent the range of answer-variable values for a given time period. A time-series plot itself, however, actually has no predisposition toward delineating time on the X axis. That is, the tick marks along the X axis need not represent segmented time.

For example, consider the corporation with seven operating business units. Corporate management might be interested in comparing the net incomes from each unit for a given year. A time-series plot can be constructed in which the X-axis tick marks represent the names of business groups while the Y axis, in this case, represents net income in dollars. Such a plot appears in Figure 12.4.

Similarly, a law firm might wish to compare strategies concerning a potential litigation situation. In this case, X-axis tick marks could represent options such as settle, arbitration, adjudication, jury trial, etc.

SOME THINGS TO CONSIDER

Time-series analysis is a powerful analytical technique and is an oft-employed process in Monte-Carlo-based risk models. The marriage, however, can result in some algorithmic behaviors of which you should be aware.

FIGURE 12.3 Time-series plot with "expanded" single period.

FIGURE 12.4   Time-series plot in which X-axis entities are not time periods.

In traditional Monte Carlo analysis, each distribution representing a risk-model variable is sampled once on each iteration. Classic Monte Carlo analysis would, on each iteration, combine the selected values according to the equations to arrive at an iteration-specific answer.

The introduction of time-series analysis complicates this simple scheme. If, for example, we decide to calculate NPV based on a 5-year projection of cash flows, then on each iteration the set of equations would be executed five times — once for each year.

On the face of it, this seems reasonable and seems to introduce only a slight wrinkle in the fabric of traditional Monte Carlo processes. However, consider the following situation. We have projected margins for the next 5 years. We expect our margin to grow over the years by 10–20%. This growth is represented by a growth distribution that ranges from 0.1 to 0.2. Multiplication of the year's margin value by a randomly selected value from the growth distribution will cause our margin to increase by between 10–20% — just what we wanted. Or is it?

Recall that each distribution is sampled once on each iteration. Remember also that on each iteration when we employ time-series analysis, we apply the randomly selected values in each period of the time series — in this case, in each of 5 years. So, if on iteration 1 we select a growth rate of 15%, we will grow our margins at 15% every year. Is this what you meant and is this what happens in the real world?

Real-world yearly growth is somewhat erratic. That is, margins may grow by 12% this year, by 17% next year, by 10% the following year, and so on. If we want our risk model to emulate the real world, then we need to employ a mechanism that allows the sampling of distributions not only by iteration, but by time-series period. The risk assessment software I employ facilitates this type of period-based sampling and analysis. You should also select a software package that accommodates this process.

Well, so what if I can't or don't resample the growth distribution on each iteration? What possible difference could it make? As it turns out, a big one.

Consider the situation in which we do not resample the growth distribution across periods. In a many-iteration model (it is not uncommon to run such models for thousands of iterations), it is statistically safe to say that in a distribution that ranges from 0.1 to 0.2, on at least one iteration a value very near 0.2 will likely be selected. Likewise, it is likely that a value at or near 0.1 will be selected on another iteration. Because these values will be applied in each year, the resulting year-by-year plot of margins will have a much greater range (for each year) than would a plot resulting from period-by-period sampling of growth. In the single-selection case in which 0.2 was selected as the growth rate, each year will grow by this amount. This will generate a significantly "higher" margin upper bound than if the growth rate were resampled for each period. The value of 0.2 might be selected for year 1, but lesser values are likely to be selected for subsequent years. This will significantly lower the upper bound for margin projections. Likewise, selection of 0.1 and employment of that value across all periods will give an unrealistically pessimistic view of margin growth.

Well, having been convinced that this is a real and acute problem, we have purchased or built a risk system that can accommodate period-based sampling of distributions. So now we've got it licked, right? Wrong. There is a little thing called dependence that has to be considered.

Traditionally, when dependence between two variables is established, one variable is designated the independent variable and the other the dependent variable. This Monte Carlo process first randomly samples the independent variable. Then, depending on the degree of correlation between the two variables, the process establishes a dependent-variable value that is reasonable for the independent-variable coefficient. This process happens once for each iteration.

Well, now we are resampling our distributions not only for each iteration, but across N periods within an iteration. If the distribution being resampled period by period has dependence relationships with the other variables, then those relationships need to be honored across periods, not just across iterations. Again, a risk-model-building software package should be purchased or built that will accommodate such a process.

DEPENDENCE — WHAT IS IT?

In the real world, parameters that are co-contributors to a risk assessment may be independent. For example, as part of a total income calculation a risk model might contain two distributions for the variables "Number of States in Which We Will Operate in the Year 2000" and "Number Of Countries in Which We Will Operate in the Year 2000." In this business, the domestic and international arms of the corporation are separately managed and funded. In the risk model, then, the afore-mentioned distributions can be independently sampled and no "invalid" pairs of values can be generated (i.e., in the risk model, any number of states can be combined with any number of foreign countries).

In an actual risk model, many variables exhibit dependent relationships. For example, in the risk model we use in the domestic part of the corporation, our income calculation considers variables such as Stage of Technology and Customer Commitment. For both variables, we might generate a distribution. In the risk model, we would not want to independently sample these two distributions because such sampling might combine, in a single Monte Carlo iteration, high Stage of Technology values with unrealistically low values for Customer Commitment. These two variables have a "relationship" in that as our corporation is increasingly viewed as the leader in our technical field, the commitment of customers (percentage of customer base, number of long-term and binding contracts, etc.) increases. The concept of dependence is that of honoring the relationship between two for more variables.

INDEPENDENT AND DEPENDENT VARIABLES

To invoke the concept and underlying technology of dependence, the first item on the agenda is to decide, for a pair of variables exhibiting dependence, which is the independent and which is the dependent variable. That is, which variable controls the other?

In the case of the Stage Of Technology and Customer Commitment variables, it may be clear in our corporation that our customer commitment increases as a result of our technical prowess (as opposed to our technical ability increasing because we are exposed to more and varied customers). In a different company, therefore, it might be that increased technical ability results from the diversity and size of the customer base.

In our situation, Stage of Technology would be considered the independent variable and its distribution would be randomly sampled by the Monte Carlo process. The value selected from the Customer Commitment distribution would, on each iteration, depend upon the Stage of Technology value selected. Thus, Customer Commitment is the dependent variable.

For cases in which it cannot be easily discerned which is the independent (controlling) and which the dependent (controlled) variable, the independent variable should be selected based upon your confidence in a given variable. For example, in another company, it may not be clear which of our two variables (Stage of Technology or Customer Commitment) is the controlling entity. They may know, however, that the values in our Customer Commitment distribution were gleaned from a comprehensive, formatted, and rigorous survey of all customers. Values that comprise the distribution for Stage of Technology were generated in a meeting of managers, many of whom were are not intimately familiar with the technology. In this case, they will have much more confidence that the Customer Commitment values actually represent reality and, therefore, the Customer Commitment variable would be selected as the independent variable.

DEGREE OF DEPENDENCE

When it has been established in a risk assessment that we indeed have a dependent relationship between two or more variables, and when we have decided which of

Variable X

FIGURE 12.5 Cross plot of two independent (uncorrelated) variables.

the variables will serve as the independent variable and which as the dependent parameter(s), the next hurdle is to establish the strength of the relationship (dependence) between the variables. The strength of a relationship can be calculated or estimated in several ways.

The strength of a dependence usually is expressed as the degree of correlation between two variables (see Selected Readings for this chapter for references relating to correlation). Correlation can loosely be described as the degree to which one variable "tracks" another. Correlation is usually expressed on a scale of –1 to +1, with 0 indicating no correlation at all (no relationship between the variables).

Figure 12.5 is a graphical representation of two variables that are independent; i.e., they are uncorrelated and have a correlation coefficient of 0. Two such variables might be the number of eggs used on a day of business in the corporate cafeteria and the number of cars that passed by the building that day. Because the number of eggs used and the number of cars passing by have nothing to do with one another, for any number of cars (any X-axis value), any number of eggs (any X-axis value) can be chosen. So, knowing how many cars passed by on a given day does not help at all in predicting how many eggs were used on that same day. There is no correlation and, therefore, a correlation coefficient of 0 is appropriate.

The opposite end of the correlation spectrum is perfect correlation, indicated by a correlation coefficient of –1 or +1. A –1 value indicates perfect negative correlation and +1 a perfect positive relationship. Well, what does that mean?

The minus (–) part of the –1 correlation coefficient means that as the value of one variable increases, the value of the other variable decreases. The 1 means that the correlation is without variance. A good real-world example is the gumball machine in the company lounge. Concerning the gumball machine, we may record how many gumballs are in the machine each day (assuming it is not refilled) and

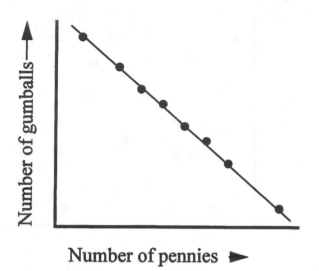

FIGURE 12.6 Cross plot of two variables with a correlation coefficient of near –1.

how many pennies are in the machine (I will here give some indication of my age by assuming that each gumball costs 1 penny — oh, the good old days!). After recording the number of gumballs and the number of pennies each day for 6 months, we plot the data and find that it looks like the plot in Figure 12.6.

From this plot, we find that if we know the number of pennies, we can predict exactly (correlation of 1) the number of gumballs left. We also know that this is a negative relationship (the "–" part of –1), because as the number of pennies goes up, the number of gumballs goes down.

In Figure 12.7 is shown a plot of perfect (1) positive (+) correlation. This is a plot from Corporate Security which indicates what they wish was true. The plot shows the difference between the number of access-card readings from people entering the building less the number of cards read when people left the building plotted against the number of people in the building. According to Security theory, the number of cards read for people coming in less the number of cards read for people exiting should exactly match the number of individuals in the building at any given time. So, knowing the difference in card readings (X-axis values) should exactly predict the number of people (Y-axis values). This should be a perfect (1) positive (as values for one variable increase, values for the other variable increase) relationship.

The fire marshal has been to visit us and has noted that we have only two fire doors. She believes that we have more people in the building than only two doors can handle and, thus, thinks we need at least a third door. Corporate Security has been charged with counting noses.

Security thought that they could tell exactly how many people were in the building by examining their card-reading data. However, several counts of bodies revealed that the actual number of people in the building was not exactly predicted

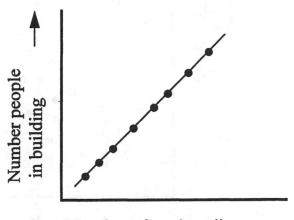

FIGURE 12.7   Cross plot of two variables with a correlation coefficient of near +1.

by the card reading data. They discovered that it is typical, on both entering and leaving the building, that one person will open the door with their card and several people will enter or exit while the door is open.

Therefore, while it is true that the number of card readings is some indication of the number of people in the building, it is not the perfect predictor (+1) they would like. So, if the number of cards read does give them some indication of body count, the correlation between the two variables is not 0. Likewise, it is not +1 because it is not a perfect predictor. It follows, then, that the relationship between these two variables should be indicated by a value somewhere between 0 and +1.

If historical data are available (as in the case where Corporate Security has card reading and actual body count data), the typical method employed to establish a relationship strength is to load the data into a software package that will perform a linear (or nonlinear if you want to get really fancy) regression. Linear regression models will return an r-squared value (correlation coefficient) that will give a reasonable estimate (for linearly varying data) of the strength of the correlation or relationship between the two variables. That is, how good a predictor of the dependent variable (body count) is the independent variable (card readings)?

Again, 0 indicates that it is no good at all; 1 indicates that it is perfect. Neither of these is true in this case. Therefore, a plot of card readings vs. number of bodies would look something like the plot in Figure 12.8. A regression model run on these data might return a correlation coefficient of, say, 0.8. This indicates that for any value of card readings selected, there is some reasonable range (Y1 to Y2 in Figure 12.8) of Y-axis values (number of people in the building).

Remember, a correlation of 1 yields a single point on the Y axis for any X-axis value selected. A correlation of 0 means that for any X-axis value selected, any Y-axis value is valid. So as we move from 1 to 0, the range of Y-axis values that would be reasonable "matches" for a given X-axis value gets greater and greater until, at 0, the entire range of Y-axis values is available.

FIGURE 12.8 Cross plot of two variables with a less-than-perfect correlation. Range Y1-Y2 represents range of reasonable Y values for the randomly selected X value.

So if we determine that there is a dependence between two variables, and if we then decide which variable is independent and which dependent, and if we assign, based on historical data and regression analysis, a correlation coefficient to the relationship, then the dependent variable will "track" the independent variable. That is, on each iteration of Monte Carlo analysis, the independent variable will be randomly sampled, and the value for the dependent variable will be randomly selected from a restricted range (like that of Y1 and Y2 in Figure 12.8) within its distribution.

When historical data are not available and when you know that two variables are correlated, an educated guess will have to be taken as to the magnitude of the correlation coefficient. This is not as bad as it may sound. First, it is most important that two correlated variables are made to "track" one another. Often, the exact (if there is such a thing) calculation of a correlation coefficient is less important. In my classes I generally have students first assign a "strong," "moderate," or "weak" label to each variable pair that exhibits some correlation. A weak relationship can be represented by a correlation coefficient between something greater than 0 and ±0.55, a moderate relationship by a coefficient between ±0.55 and ±0.75, and a strong relationship by a coefficient greater than 0.75. This certainly is not an exact science, but it generally serves the purpose.

MULTIPLE DEPENDENCIES AND
CIRCULAR DEPENDENCE

In any given risk assessment there can be many pairs of variables that exhibit dependence. For example, the number of employees might correlate with the number of sick days taken (positive relationship — as one goes up, the other goes up) and the net income for our business unit may correlate with our costs/expenses (negative relationship — as costs go up, income goes down), etc. It may also be true that one independent variable can have several associated dependent variables. Going back to our Stage of Technology example, it may be that as an independent variable, Stage of Technology may "control" several dependent variables. We already discussed the fact that at our corporation, as Stage of Technology increases, our Customer Commitment increases. This, then, is a positive correlation and would be indicated by a correlation coefficient between 0 and 1. However, it may also be true that our Stage of Technology also affects our corporation's cash exposure. That is, as we are increasingly viewed on Wall Street as the leader in this technology, it is increasingly easy for us to attract investors, thus lowering our corporation's cash investment. So, as Stage of Technology increases, corporate cash exposure (investment) decreases. This is a negative relationship that would be represented by a correlation coefficient between 0 and −1.

We can set up as many variables to be dependent on a single variable as we like (i.e., as many children dependent on a parent as we like). In this scenario, though, one must be careful not to set up circular dependence.

Any software package worth the CD it comes on should automatically guard against circular dependence. For example, for three variables, A, B, and C, circular dependence is employed if I say A is dependent on B, and B is dependent on C, and C is dependent on A. This circular logic should be rejected by the system. In this simple example, the circular reasoning is obvious. However, in complex technical and economic risk assessment models, it can be easy to inadvertently attempt to invoke such relationships.

EFFECT OF DEPENDENCE ON MONTE CARLO OUTPUT

Consider two independent (correlation of 0) distributions A and B from which we will calculate C by the formula $C = A + B$. The distribution for variable A ranges from 2 to 10 with a most-likely value of 4. Distribution B ranges from 20 to 30 with a most-likely value of 24. Both distributions are shown in Figure 12.9.

On a given Monte Carlo iteration, if an extreme value of A is selected (say, a value of 9), it most likely will be combined with a near-most-likely value from B. The resulting cumulative frequency plot for C would look like that in Figure 12.10 and might range from a minimum C value of 23.7 to a maximum C value of 36.4.

If, however, a strong dependence of, say, 0.9 is established between A and B, then on any given Monte Carlo iteration when an extreme value of A is selected, dependence will "force" an extreme value of B to be combined with that of A. This will result in the plot of C shown in Figure 12.11, in which C now ranges from 22.4 to 40.0. A major effect, therefore, of implementing dependence is to change (generally expand) the range of the "answer" variable.

DEPENDENCE — IT'S UBIQUITOUS

Whenever I offer a risk class for managers, the class is given *sans* computers (some managers may read this, so I won't go into *why* we don't use computers). In a "computerless" environment, exercises must be designed to bring home a point

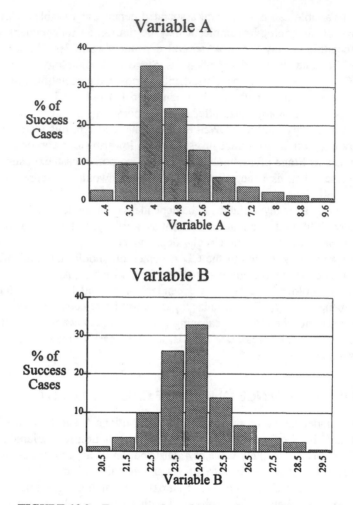

FIGURE 12.9   Frequency plots of independent variables A and B.

without venturing too far into the erudite technical haze. One of my favorite com-
puter-free exercises is one I use to illustrate the ubiquitous nature of dependence
and the need to consider it.

In Figure 12.12 is a list of variables from an actual plant-construction model.
The original list was much longer. The list presented here has been whittled down
for simplicity.

I typically challenge the class to peruse the list and then discuss which pairs of
items on the list have real-world relationships (i.e., dependencies). I always get them
kick-started by offering an example. The suggestion is put forth that Public Perception and Litigation Costs are linked, the reasoning being that if you are perceived
as an "evil empire" by the public, you will spend more time in court. I then draw
a line connecting these two items indicating that they have a relationship. As public

FIGURE 12.10 Cumulative frequency plot resulting from using Monte Carlo analysis to combine two independent variables A and B.

FIGURE 12.11 Cumulative frequency plot resulting from using Monte Carlo analysis to combine two highly correlated (0.9) variables A and B.

perception of your company worsens, litigation costs go up. I then challenge the group to come up with some connections of their own. Typically, some lively discussion ensues.

Dependence

Public perceptions
Market research
Cost analysis
Production capacity
Demand
Profit margin per unit
Conversion costs (depreciated)
Conversion costs (actual)
Advertising costs
Litigation costs
Operating revenue
Tax rate
Marketing synergies
Taxable income

FIGURE 12.12　List of variables in dependence exercise.

The number of lines connecting variables in Figure 12.13 is about half of what can be drawn (I excluded the rest of the lines because it just makes too big a mess). As can be seen, there are many relationships, that is, dependent-variable pairs. In class, I project the original list on a white board or flipchart on which I draw the connecting lines. When the projector is turned off, you are left with just the tangle of lines shown in Figure 12.14 (remember, only about half the lines are shown in Figure 12.14). Class members generally are impressed concerning the number of relationships and by the fact that there is nary a variable on the list that can be independently sampled — that is, sampled without consideration of at least one other variable.

I make the point that if you wish your risk model to emulate the real world, then the relationships shown must be honored. Even using such a simple list of variables, the relationships can be many and complex. If not considered, your risk model will be emulating something, but likely not what you intended.

Dependence

FIGURE 12.13   List of variables with dependencies shown as connecting lines.

Dependence

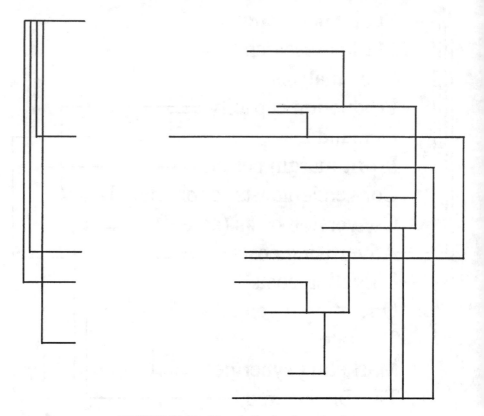

FIGURE 12.14   Lines indicating dependencies.

13 Risk-Weighted Values and Sensitivity Analysis

CONTENTS

Introduction to Risk-Weighted Values and Sensitivity Analysis 159
Risk-Weighted Values — Why? 159
Risk-Weighted Values — How? 161
The Net Risk-Weighted Value 164
The Economic Risk-Weighted Resource (ERWR) Value 165
Risk-Weighted Values — The Answer 166
Sensitivity Analysis — Why? 168
Sensitivity Analysis — How? 170
Sensitivity Analysis — Results 172

INTRODUCTION TO RISK-WEIGHTED VALUES AND SENSITIVITY ANALYSIS

When a risk assessment is completed, interpretation of the results is the next step. Making this task as effortless as possible can be critical to the success of a risk model. If output from a risk assessment can't be understood by those to whom it is presented, it likely was a wasted effort to have pursued the risk analysis in the first place.

Risk-weighted values facilitate the integration of several risk-model-output components. The various types of risk-weighted values make simple the interpretation and comparison of output from multiple assessments.

Likewise, sensitivity analysis aids in identifying the elements of a risk model that were most and least important to the calculation of the answer. Following a risk-model run, examination of a sensitivity plot can be a fundamental step in deciding what action to take as a result of the analysis.

RISK-WEIGHTED VALUES — WHY?

The reason we need risk-weighted values always reminds me of a dog chasing a car. The dog races down the street, expending copious quantities of energy, and finally catches the car. The question now becomes, "OK, buddy, you caught it — now what are you going to do with it?"

So it is with risk assessments and their output. We struggle arduously to generate a detailed, comprehensive, and practical risk model and its resulting plots and erudite statistics. Well, OK, buddy, you got the results, now how are you going to interpret and compare output from multiple risk model runs?

Investment A

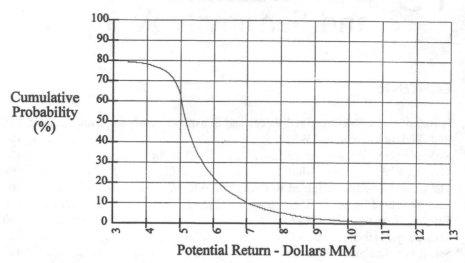

FIGURE 13.1   Cumulative frequency plot for Investment A.

There is use in assessing the magnitude of benefit or cost and the related probabilities associated with a single and isolated event. Relatively large returns, however, only are realized when we undertake the comparison of multiple scenarios or events. Without employing risk-weighted values, the process of comparison of risk-model outputs can be difficult and at times practically impossible.

Consider two investment opportunities, A and B. Opportunity A has a 20% chance of falling through. That is, if we invest in opportunity A, we have a 20% chance of losing our investment and realizing no return. If opportunity A does not fail, it will return no less than $3 million, no more than $12 million, and will most likely return about $5.3 million.

Like investment A, investment B has a chance of failure. There is a 35% chance that opportunity B will fail and our investment lost. If investment B does not fail, it will return a minimum of $1 million, a most likely of about $6.8 million, and a maximum of $10 million. Cumulative frequency plots for opportunities A and B appear in Figures 13.1 and 13.2.

Inspection of the plots in Figures 13.1 and 13.2 shows that investment A has a 15% greater chance of succeeding than does investment B (20% vs. 35% chance of failure). If it does not fail, investment A has a higher minimum yield than does B. Likewise, A has a higher maximum yield. On the face of it, then, it might seem that A is the better opportunity. But wait, there's more!

From the cumulative frequency plot of investment B we can see that we have about a 45% chance of getting a return of around $6 million. The probability of realizing a $6 million return from A is less than half that of investment B. Hmm, maybe we should take a look at this in a different space.

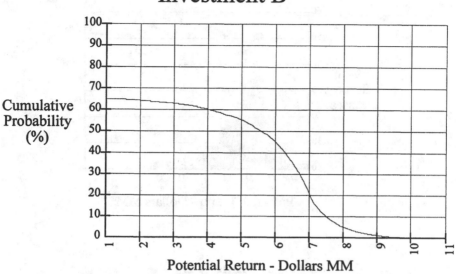

FIGURE 13.2 Cumulative frequency plot for Investment B.

In Figure 13.3 are the frequency plots for the two opportunities. Inspection of the plots indicates that the most likely return from investment A, if it does not fail, is about $5.25 million. The most likely return from investment B, if it does not fail, is about $6.8 million. Well, now which investment appears to be the better?

Risk-weighted values facilitate the comparison and ranking of probabilistically assessed opportunities. However, before we use risk-weighted values to select the best opportunity, let's take a look at just how risk-weighted values are calculated. The answer to this problem is given in the section of this chapter entitled Risk-Weighted Values — The Answer.

RISK-WEIGHTED VALUES — HOW?

The fundamental premise behind the calculation of risk-weighted values is simple. In previous sections of this book we have described just how cumulative frequency plots are constructed (see Chapter 9). In Figure 13.4 is shown a set of values that comprise a cumulative frequency curve. Also shown in Figure 13.4 is a "blow up" of the section of the cumulative frequency curve containing the two points whose coefficients are of greatest magnitude.

The X-axis difference between the two points is 2 (99 – 97). The probability of random selection of the point whose coefficient is of greatest magnitude is one chance out of however many discrete X-axis values there are to choose from (the number of X-axis values is equal to the number of Monte Carlo iterations). Therefore, in the construction of the cumulative frequency plot, the "last" point is elevated one

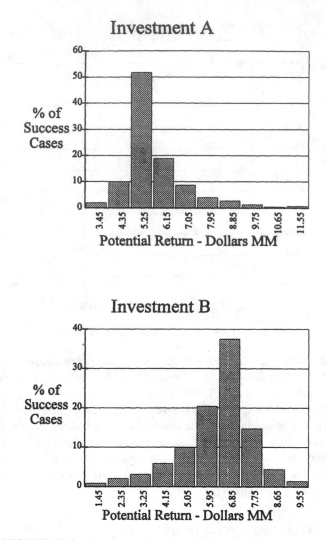

FIGURE 13.3 Frequency plots for two investment opportunities.

probability distance above the X axis. The Y axis is subdivided into equal-sized pieces. The number of pieces equals the number of Monte Carlo iterations. One "piece" is equal to one probability distance along the Y axis. The data value whose coefficient is second in magnitude is elevated two probability distances above the X axis (in a cumulative frequency plot, there are two chances out of the total number of discrete values of selecting the point of second-greatest magnitude *or* the point of greatest magnitude).

So, we know that the vertical legs of our trapezoid are probabilities and that the base of the trapezoid is the X-axis difference between the two points. We can now calculate the area of this trapezoid and all other trapezoids to the "left" of this one. The sum of all the individual trapezoidal areas is the risk-weighted value (we will

Risk-weighted values – how?

A cumulative frequency curve, although usually represented as a smooth line, is really a series of plotted points which represent the sorted values resulting from Monte Carlo analysis. The number of points is equal to the number of iterations in the Monte Carlo analysis.

The area of the trapezoid is calculated using the probabilities associated with each point as the vertical legs of the trapezoid and the difference between the values as the base. The risk-weighted value is simply the sum of all trapezoids formed by all pairs of adjacent points.

Expanded area

Probability associated with value of largest magnitude

Probability associated with value of 2nd-largest magnitude

Difference between values = 2

95

97

100

100

FIGURE 13.4 Plot showing how a risk-weighted value is calculated from a cumulative frequency curve.

see later that if the cumulative frequency plot reaches 100 on the Y axis, then the risk-weighted value so calculated will equal the mean of the X-axis values).

This methodology, in effect, calculates the area under the curve. Calculation of this area, however, is generally unsatisfactory for many reasons, not the least of which is the fact that the area remains constant regardless of where the curve is situated on the X axis. That is, the risk-weighted value for a curve with a minimum of, say, 10 and a maximum of 100 is exactly the same as the same curve "slid down" the X axis so that its minimum is 110 and its maximum is 200. The magnitude of the risk-weighted value does not reflect the position of the curve on the X axis. Therefore, some customized risk-weighted values generally are calculated. We will consider two of these.

THE NET RISK-WEIGHTED VALUE

No matter whether a risk assessment begins as a technical, political, environmental, or other type of evaluation, typically the result of such assessments is a monetary or financial calculation. A most useful parameter for comparing and ranking financial risk model output is the "net" risk-weighted value.

In the previous section it was explained how, in principle, a risk-weighted value is calculated and that calculation of only the area under the curve is not an intrinsically utilitarian measure. If risk-weighted values are to be used to compare and rank multiple risk assessed opportunities, then a mechanism must be employed whereby the changing position of the cumulative frequency curve, relative to the X axis, invokes a change in the risk-weighted value.

For the net risk-weighted value and the economic-risk-weighted-resource value (to be discussed in a subsequent section of this chapter), we introduce the concept of the "limit line." In Figure 13.5 a limit line has been selected that is coincident with the Y axis and the X-axis value of 0. The cumulative frequency curve in this figure is typical of many such curves that result from financial risk assessments in that the curve "straddles" 0. That is, the curve represents both positive (making money) and negative (losing money) values. X-axis values that plot far from 0 in either the positive or negative direction are of relatively great magnitude. The probabilities associated with these values, however, are relatively small. Therefore, on the negative side of the cumulative frequency curve, like their positive counterparts, data points that plot increasingly far from 0 represent increasingly "large" negative values which also are increasingly less likely to happen in the real world (increasingly small associated probabilities).

As depicted in Figure 13.5, the "positive" area under the curve is calculated. This is done in the manner previously described. The negative area is calculated in exactly the same way (on the negative side, our trapezoids would be "upside down" with respect to their positive-curve counterparts). The negative area is then subtracted from the positive area to yield a net risk-weighted area. This risk-weighted measure is a favorite tool of economists and others who must evaluate, compare, and rank financial opportunities. An example of its use will be given at the end of this chapter.

A net risk-weighted value, like any other risk-weighted value, is not an indication of how much an opportunity is expected to produce. That is, the risk-weighted value

Before-Tax Annual Increase in Cash Flow: 1999

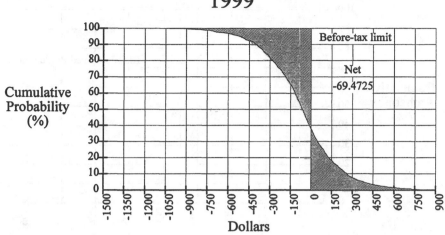

FIGURE 13.5 Cumulative frequency plot. Shaded areas are used to calculate net risk-weighted value.

is not equal to (necessarily) the modal or "most likely" value on the answer-variable plot. Rather, a stand-alone, risk-weighted value indicates the amount of dollars (or whatever the X-axis represents) that the risked opportunity would contribute to a portfolio of opportunities on a fully risk-weighted basis. The best use of a risk-weighted value, however, is as a means of comparison of one opportunity to the next. They are most commonly used as a relative measure. See the section Risk-Weighted Values — The Answer of this chapter for an example.

THE ECONOMIC RISK-WEIGHTED RESOURCE (ERWR) VALUE

When the answer variable is expressed in other-than-monetary units such as gallons or barrels of liquid, tons, or other units that generally do not "go negative," the economic risk-weighted resource (ERWR) value may be employed. Like the net risk-weighted value, calculation of the ERWR utilizes a limit value — most commonly an economic limit.

Consider the situation in which we are in the midst of a labor strike. We have only enough nonunion workers to operate one of our four factories. In Figure 13.6 is depicted a cumulative frequency curve that represents next month's projected production, in tons, from one of the factories. Values on the X axis indicate that next month's total production from the factory being considered for operation could range from 0 tons (if we, in fact, can't make it work without union labor help) to 1000 tons. Our economists indicate that if total production is not at least 450 tons next month, then we will lose money for the month.

In response to our economist's calculated lower acceptable limit of production, we have drawn a limit line on the plot at 450 tons. The question we would like

Projected Production

FIGURE 13.6 Cumulative frequency plot of projected production with limit line drawn at 450.

answered is, "What is the probability of achieving our lower limit of production, and what is the ERWR value (to be used in comparing one factory's projected output to that of the three others so we can decide which factory to operate)?"

An ERWR value is determined by calculating the area under the cumulative frequency curve "up to" the limit line and adding to this area the area of the "box" to the "left" of the limit line. The "box" extends all the way to the X-axis value of 0 whether or not it is shown on the plot. The shaded area in Figure 13.7 indicates the ERWR area. This, at first glance (or even at second glance), may seem to be a queer area to serve as the basis for a risk-weighted value.

Figure 13.8 is the frequency plot for the same data. The area represented by the bars to the "right" of the limit line in Figure 13.8 is the same area as that shaded in Figure 13.7. The area to the "left" of the limit line in Figure 13.8 is equivalent to the unshaded area in Figure 13.7 under the cumulative frequency curve. As with the net risk-weighted value, the ERWR is not the most likely amount the factory will produce next month (most likely is somewhere in the vicinity of the tallest bar in Figure 13.8). Rather, it is the amount of production that this factory would contribute to a portfolio of outputs of factories on a fully risk-weighted basis. In the section immediately following, an example is given that makes clear the use of risk-weighted values as ranking tools.

RISK-WEIGHTED VALUES — THE ANSWER

As indicated in the immediately preceding sections of this chapter, individual risk-weighted values do have meaning in the context of portfolio analysis. The real power of risk-weighted values, however, is mainly realized when they are used to compare opportunities.

Projected Production

FIGURE 13.7 Cumulative frequency plot of projected production. Shaded area is area used to calculate ERWR (risk-weighted production) value.

Projected Production

FIGURE 13.8 Frequency plot equivalent to the cumulative frequency plot in Figure 13.7.

Let's return to the original problem presented in the Risk-Weighted Values — Why? section of this chapter. This problem presents us with the task of choosing between two investment opportunities. Probabilistic analysis of the two opportunities

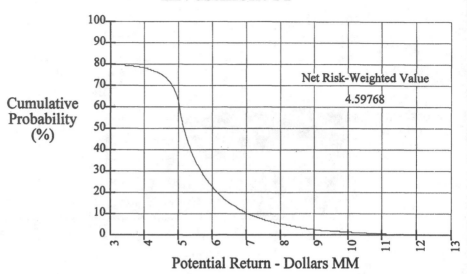

FIGURE 13.9 Cumulative frequency plot for investment A and net risk-weighted value.

yields the plots shown in Figures 13.1 and 13.2. The X axes of the cumulative frequency plots for both opportunities cover different ranges. The curve shapes are different. The Y-axis intercepts (chance of failure) of the two investments are not equal. Other parameters such as economic limits could be different (but are not in this example). How does a person integrate in their minds changing ranges, changing curves, changing intercepts, etc.? Such integration is exactly the service provided by risk-weighted values.

Because investments generally are considered relative to 0, the limit lines used in the calculation of the net risk-weighted values in Figures 13.9 and 13.10 are at 0. The net risk-weighted value for investment A is 4.597. That for investment B is 4.125. Because the X axes of these plots represent dollars to be gained, then investment A is the best choice (all other things being equal, of course). If the X axis represented dollars to be paid, such as costs, then the smaller value would be the best choice.

SENSITIVITY ANALYSIS — WHY?

The purpose of most risk analyses is to ascertain the validity and quality of one or more opportunities. Most comprehensive risk studies are composed of many, sometimes dozens, of input and output variables. These variables are linked by a system of equations that typically integrate technical, political, financial/commercial, and other parameters.

Each opportunity is assessed by the risk model with the expectation that output from the model will reveal the opportunity to be of good or poor quality — that is,

FIGURE 13.10 Cumulative frequency plot for investment B and net risk-weighted value.

an opportunity you might pursue or one you should forego. If risk-model results indicate that a particular opportunity is of high quality, you should endeavor to determine which risk-model input parameters contribute most to its relative stellar standing. For example, it may turn out that an investment opportunity ranks high primarily because of the quality of the labor force (as opposed to favorable interest rates, quality raw materials, or other risk-model input parameters). Steps should then be taken to preserve the quality of the labor force so that the quality of the opportunity can be maintained. This is not to say that other parameters also are not important; however, there generally is only so much money, time, and effort that can be directed toward any given opportunity. Therefore, it is essential that we discover and nurture the most important "contributing" variables.

Conversely, the assessment of an opportunity may reveal it to be of poor quality. However, for political or other reasons we may wish to pursue the opportunity if its quality can be improved. A given opportunity may contain political, financial, environmental, commercial, technical, and other input variables. These parameters typically serve as the terms in complex equations that integrate them. If the quality of an opportunity is to be enhanced, it is essential that we discover which parameters, if "improved," would result in the greatest increase in quality of the opportunity. With limited resources it is beyond our ability to "fix" everything, so it is essential that we attempt to "fix" the parameters that will most improve the overall quality of the opportunity.

Inspection of the system of equations that comprise a risk model usually is of little or no help when attempting to identify critical variables. Unless the equations are simple sums or other equally credulous notation, perusal of the lines of computer code ordinarily will not reveal the magnitude of influence associated with individual variables. Neither can such influence, again with the exception of the most simple equations, be deduced from the magnitude of the coefficients which represent the variables. For example, if in a Monte Carlo-based risk model we consider the simple equation

$$C = A + B \tag{13.1}$$

in which A is a distribution with a minimum of 10,000,000 and a maximum of 100,000,000 while B is represented by coefficients that range from 2 to 6, it should be obvious that the value of C is overwhelmingly influenced by variable A. Variable B exhibits little influence. Variable A, therefore, is the "controlling" parameter. However, if we change the equation to

$$C = A^B \tag{13.2}$$

now it is clear that the influence of B on the value of C has dramatically increased even though the magnitude of the coefficients representing A and B are unchanged. Therefore, except in the most mundane of circumstances, it is folly to suppose that the variables whose coefficients are of greatest magnitude also are those which exert the greatest influence on the "answer" variables.

SENSITIVITY ANALYSIS — HOW?

Through the years, many attempts have been made to develop algorithms which will, for a complex set of equations, reveal which of many input variables are those of greatest influence on the output variables. One of the more popular but significantly flawed methodologies is to hold all variables, save one, at a constant value in order to determine what influence the variable not held constant has on the coefficients of the output variables. It should not take a rocket scientist, however, to realize that the true influence on an output variable depends on the interaction of all coefficient ranges for all variables and that the "hold-all-but-one-constant" method might not, in most cases, reveal the true influence exerted by any variable. Given the popularity of this approach, it is worth a few paragraphs here to describe some of the drawbacks to the "hold-all-but-one-constant" (HABOC) method.

First, I would like to state, and will, that there is nothing intrinsically amiss with the form of a tornado diagram. If, for example, we have measured or calculated the ranges of net incomes of various businesses, there is not a thing wrong with displaying those net-income ranges as a tornado plot. The range in net income for each business can be represented by a bar of appropriate width and placed on the diagram accordingly. Such tornado plots are informative and aesthetically satisfying. So, the question concerning tornado diagrams is not whether they are truthful or useful, but rather how the information displayed was calculated.

In a risk model, the coefficients for all variables change, at least, with each iteration. It is difficult, by inspection, to ascertain the contribution of a single input variable to the change (relative to the previous iteration) in the value of an output variable. To remedy this, the HABOC methodology was born.

Using this approach, in order to determine the influence of a single input variable on an output variable, all variables, save one, are held at a constant value. The value used as the constant typically is a value determined to be "representative" of the distribution. Mean values, median values, and those termed "deterministic" commonly are employed. The Monte Carlo process is imposed, holding all variables constant save one that is allowed to randomly vary throughout its range. Each variable has its turn to be allowed to vary while all others are held constant. For each variable allowed to change, the range of the output variable is noted, and this becomes the "width" of the bar for that input variable in the tornado plot. For speed (or for lack of sophistication), some approaches use only, for example, the 10th, 50th, and 90th percentile values for the variable that is allowed to change value. Discretization only worsens the problem.

The first thing that should strike you as suspicious concerning this methodology is that it violates dependence (see Chapter 12) and creates illegal combinations. For example, consider the two variables, Depth and Temperature, in a risk model. In the real geological world, these two variables are related. As depth (in the earth) increases, so does temperature (the deeper you go, the hotter it gets). This relationship is termed the "geothermal gradient" and varies with geographic location.

The Depth distribution might contain values ranging from 0 feet to 20,000 feet. Concomitant temperatures might range from 50°F to 500°F. Using the HABOC method, if you are attempting to determine the influence of Depth on an answer variable, you would allow Depth in the Monte Carlo process to vary throughout its 0 to 20,000 range while holding Temperature constant at its "deterministic" value. Because I have not here displayed either the Depth or the Temperature distributions, allow me to suggest that the "deterministic" value for Temperature is 220°F (it does not matter what number you choose as the constant Temperature value — the result is the same).

Using the HABOC technique, Depth is allowed to vary while Temperature is held constant. So, Depth values of 100 feet are combined in the equations with the constant 220°F Temperature as are Depth values of 19,000 feet. These and countless other Depth/Temperature pairs are, as far as nature is concerned, illegal combinations. Yet, the variation of the answer variable is still, under this scheme, considered to reflect the influence of Depth on that variable. I don't think so.

Still another and more insidious drawback to the HABOC method is linked to branching in the risk-model logic. Consider the situation in which we have two variables A and B that are combined to create an output variable C. Variable A ranges, say, from 0 to 100. Consider also that our equations for calculating C contain a probabilistic branch thus:

$$\text{If } A < 50 \text{ then} \tag{13.3}$$

$$C = B + 1$$

$$\text{Else}$$

$$C = B^{100}$$

When attempting to determine the influence of B on the answer, the HABOC logic would have us hold A constant at some "deterministic" value. Again, it is irrelevant what value we select as the constant for variable A — let's pick 55. So, on each iteration the model will execute the If/Else logic and always calculate C to be equal to B^{100}. The C value of B + 1 will never be executed. The bar width indicating the influence of B on C will be drastically exaggerated because if A were allowed to vary, the equation B + 1 would be executed a statistically significant number of times. Therefore, the bar representing the influence of B on C is grossly in error. There's a better way.

To date, the most satisfying methodology developed for the purpose of discovering the relative influence of input variables upon the magnitude of coefficients for output variables is the Spearman rank correlation (SRC) method. In keeping with the practical bent of this book, a detailed description of the SRC equations was deemed unnecessary dunnage. Complete disclosure of the SRC methodology can be found in nearly any textbook on basic statistics (see Selected Readings section for this chapter). However, it would be insouciant to refrain from putting forth at least a cursory description of Spearman's correlation method.

SRC is a nonparametric (distribution-free) technique that facilitates the calculation of correlation while allowing all parameters in a multivariate analysis to vary simultaneously. Correspondence between individual input variables and a calculated output variable is arrived at by comparison of ranks. Ranks are employed in situations in which actual measurements are impractical or impossible to obtain. A mundane example is an ice-skating contest in which judges assign subjective scores. The scores are used to rank the skaters. In a two-judge (1 and 2) and three-skater (A, B, and C) contest, judges may both rank the skaters in the order B, C, A. In this case there is perfect correspondence (+1) between the rankings. If judges 1 and 2 had ranked the skaters in exactly opposite order, the rank correlation would be –1.

If measurements rather than ranks comprise the original data set, then the measurements must be converted to ranks before SRC can be applied to the data. Conversion from measurements to ranks likely will result in the loss of some information. Therefore, it is unrealistic to expect that the product–moment correlation coefficient and that resulting from SRC will be equal when both methods are applied to a common set of data.

Loss of information notwithstanding, SRC may, relative to product-moment correlation, be a superior measure of correspondence. SRC's prelate position is derived from the fact that it is not predicated on the usually unwarranted presumption that the coefficients of the input variables are normally distributed. Correlation between variables representing coefficient distributions of nearly any type can receive valid treatment by the SRC technique.

SENSITIVITY ANALYSIS — RESULTS

Output from SRC analysis is commonly displayed as a "tornado" diagram such as that shown in Figure 13.11. On a scale of –1 to +1, the length of the bars indicates the degree of correspondence between a given input variable and the calculated output variable. A separate tornado plot must be produced for each output parameter.

FIGURE 13.11   Sensitivity plot.

In the plot shown in Figure 13.11, the numbers in parentheses adjacent to the variable names indicate the percentage of variance of the output variable that is explained or accounted for by that input variable. Bars extending to the right of the 0 center line are variables that are positively correlated with the output variable (as coefficients for the input variable increase in magnitude, so do the coefficients of the output variable). Bars extending to the left of the center line indicate negative correlations.

Sensitivity plots like the one shown in Figure 13.11 are not only fundamental to determining which are the prominent input variables, but can be invaluable indicators of whether a particular project should be pursued. For example, the results from a risk model may indicate that we likely will lose money if we pursue the risk assessed opportunity. For political reasons, however, we would like to go ahead with the project if the losses are not significant. We hope that our sensitivity analysis will reveal which of the aspects of the project we might seek to improve, thereby enhancing the palatability of the endeavor.

Our plot may indicate, however, that the two most significant variables, accounting for the majority of the variance of the output variable, are parameters over which we have little or no control. Parameters such as market price and taxes are good examples. As in this case, if parameters beyond our control are the significant factors in a discouraging risk assessment, then the sensitivity plot is not only an indicator of which variables are significant, but also is an indicator of whether it is logical and beneficial to further pursue an opportunity.

14 Risk Assessment Examples

CONTENTS

Introduction..175
Qualitative Model 175
Semi-Quantitative Model...177
Mini Time-Series Model 180
Fate-Transport (Process/Health) Model 187
Decision-Tree Conversion 192
Legal Model..198
Comprehensive Risk Assessment Example..204

INTRODUCTION

It is my intention in this chapter to present to the reader some very simple risk-model examples. The purpose of each model is to exemplify an approach that can be used to solve a problem of a specific type. The reader should keep in mind that most real-world risk models are comprised of dozens of variables and, usually, tens if not hundreds of lines of code.

It should also be noted that the examples below do not represent an exhaustive manual on how to solve such problems. The approaches exemplified here were selected for their simplicity and because they are most generic. Each risk model built presents the builder with a unique set of circumstances that must be handled in a way that is not only efficient, but that is not distasteful to the user.

QUALITATIVE MODEL

Construction of most risk assessment models requires the integration of "hard" data resulting from real-world measurements or forecasts, and "soft" data that are not expressed quantitatively. An example of a qualitative model may be one in which the answers to questions are statements from which the user can select one or more as the answer to the question.

One approach (one of many, to be sure, but the only one to be outlined here) to treating qualitative data is to rank-order the qualitative answers in increasing-risk order.

This is done so that the results of a qualitative assessment may be expressed quantitatively, or so that qualitative data may be integrated with quantitative input.

For example, a qualitative risk assessment question and associated answers might be as follows:

Has a security assessment been made in this country?

___ 1. Yes, very recently — favorable result.

___ 2. Yes, but some number of years ago — favorable result.

___ 3. No, but the country has no history of security problems.

___ 4. No, and the country has a history of security problems.

___ 5. Yes, and there was indication of security problems.

Because the answers to questions are statements and not numbers, it generally is good practice to convert the statements of increasing risk to a quantitative scale such as 0 to 100. For the question above, there are five answers. Given that we have decided on an output variable scale of 0 to 100 and given that there are 5 answers to this question, the first answer will be assigned a risk value of 0, the second question a value of 25, the third answer a value of 50, the fourth answer a value of 75, and the fifth answer a value of 100. The danger in this methodology is that the 0, 25, 50, 75, and 100 increments do not reflect the real-world relative degrees of risk for the answers. Weights may need to be applied to adjust each answer's risk value.

In the slots in front of each question, the user may indicate their confidence in the answer by entering a percentage value. For example, if the user thinks that some time ago there was performed a security assessment, but can find no trace of it, they may answer the question thus:

___ 1. Yes, very recently — favorable result.

_60_ 2. Yes, but some number of years ago — favorable result.

_40_ 3. No, but the country has no history of security problems.

___ 4. No, and the country has a history of security problems.

___ 5. Yes, and there was indication of security problems.

This indicates that the user is not sure whether a security assessment was done but that the country does not historically have security problems. On each Monte Carlo iteration, there will be a 60% chance of selecting answer 2 (thus a risk value of 25) and a 40% chance of selecting answer 3 (representing a risk value of 50). On each iteration the risk value from this question is combined with similar information from other questions such as the following.

What is the quality of the ground-transportation infrastructure?

___ 1. Good to major locations — paved roads.

___ 2. Roads to major locations — not paved.

_100_ 3. Poor roads everywhere.

___ 4. No usable roads.

For this question, answer 1 has a risk value of 0, answer 2 a value of 33, answer 3 a value of 66, and answer 4 a value of 100. Again, weights may need to be applied to adjust the risk values. The user has indicated that they are 100% sure that the answer to the question is answer 3. Therefore, on each Monte Carlo iteration, this question will have a risk value of 66.

On a given iteration of this two-question risk model, answer number 2 from the first question might be combined with answer 3 from the second question, resulting in a single-iteration overall risk value calculated thus:

$$\text{Total Risk Value} = (25 + 66)/2 = 45.5 \qquad (14.1)$$

where 2 is the number of questions in the risk model.

Most risk models are composed of a dozen or more questions, each of which can have a unique number of answers. In addition, weights are generally employed, in the form of decimal percentages, to emphasize the relative importance of one question or category of questions to another. For example, we may believe that because we have military or security forces in the area, the security concerns are less important in the country being assessed than are transportation problems. The user may be asked to assign weights (that sum to 1 or to 100) to each question or category of questions. In this case, the user may indicate

0.3 = Weight assigned to security problems;

0.7 = Weight assigned to transportation problems.

So, including the weights in the calculation, the total risk is

$$\text{Total Risk Value} = (25 \times 0.3) + (66 \times 0.7) = 53.7 \qquad (14.2)$$

Note that this is a different total risk value than the answer calculated without consideration of weights (45.5). The value of 45.5 is also calculated by using implicit equal weights for both questions (0.5 and 0.5).

Output from a many-questioned qualitative risk model typically appears as that shown in Figure 14.1. This risk output curve can stand alone or may be integrated with quantitative information in the same or from other risk models.

One word of warning about such models. This type of model can have serious problems with regard to data scales (see paper by Pariseau and Oswalt in the Selected Readings section for this chapter). Readers should be aware of problems concerning the application of ordinal data to risk processes before attempting to build such a model.

SEMI-QUANTITATIVE MODEL

When an ice skater wins a competition, it is because she or he was judged best relative to other participants. So, too, it is for the platform diver, the beauty-pageant

New-Country-Entry Evaluation

FIGURE 14.1   Cumulative frequency plot of total risk for new-country entry.

participant, and other contestants. In the aforementioned competitions, "measurement" of a thing is subjective, and such measurements generally are expressed on a scale such as 1 to 10, or 0 to 100, or some other fixed but arbitrary range.

So it is with many aspects of business that may contribute to a risk assessment. For example, we may be interested in evaluating the past performance of a company that we are considering as a partner. Our company may be interested in evaluating our potential partner with regard to many aspects, two of which might be

- Our relationship with potential partner's leadership;
- Potential partner's environmental compliance record.

Our company may decide to judge all aspects on a scale of 1 to 10; 1 indicating "good" or "low" risk and 10 indicating "poor" or "high" risk. For each question, we may ask the appropriate user (expert in the subject of the question) for a minimum, most likely, and maximum value on a 1 to 10 scale.

Our expert on the potential partner's leadership indicates that our company has had the best of relationships with the potential partner's present CEO. We have cooperated many times in the past and always to our mutual benefit and satisfaction. However, the current CEO is very near retirement and has warned of it. His successor is known to be much less favorably disposed toward our company and toward such partnerships. The transition from the present CEO to his successor could happen just before or during the formation of the partnership, or, it might not happen at all in the time of the partnership. Our expert, therefore, gives us the following values for the risk assessment:

cooperated many times in the past and always to our mutual benefit and satisfaction. However, the current CEO is very near retirement and has warned of it. His successor is known to be much less favorably disposed toward our company and toward such partnerships. The transition from the present CEO to his successor could happen just before or during the formation of the partnership, or, it might not happen at all in the time of the partnership. Our expert, therefore, gives us the following values for the risk assessment:

FIGURE 14.2 Frequency plot for evaluation of relationship with potential partner's leadership. Minimum value of 2, most-likely value of 3, maximum value of 8, and peakedness of 9.

Our relationship with potential partner's leadership.

- Minimum value = 2
- Most-likely value = 3
- Maximum value = 8

These values indicate that our expert believes that the most likely situation will be low risk (3), but it could "get nasty" (8) if the present CEO retires. The minimum value of 2 indicates that our treatment by the potential partner leadership could be slightly better than expected. We may also ask our expert to supply us with a peakedness value on a scale of 1 to 10 (see Chapter 10 for a discussion of peakedness) to indicate his confidence in his most likely value of 3 (relative to 2 and 8). The expert gives us a peakedness value of 9 indicating high confidence in his assessment. The resulting distribution for this parameter is shown in Figure 14.2.

Our expert on the potential partner's environmental track record has given us the following estimates:

Potential partner's environmental-compliance record.

- Minimum value = 3
- Most-likely value = 6
- Maximum value = 8
- Peakedness value = 5

FIGURE 14.3   Frequency plot for evaluation of potential partner's environmental record. Minimum value of 3, most-likely value of 6, maximum value of 8, and a peakedness of 5.

Our partnership will include some collaboration in foreign countries. Our potential partner has had a relatively good environmental compliance record in the U.S. (value of 3). They have, however, run into trouble in several of their non-U.S. projects. In fact, the company was required to terminate one foreign project due to environmental problems (thus, the most-likely value of 6 and the maximum value of 8). Because our company would have some control over environmental issues and because only a portion of the partnership venture would be foreign, a peakedness (confidence in the most likely value of 6) of 5 was assigned. The resulting distribution for this parameter appears in Figure 14.3.

These two distributions can be used as input to a risk model that utilizes the very simple equation

$$\text{Total Risk} = (\text{RWPP} + \text{ECR})/2 \qquad (14.3)$$

where RWPP is the distribution representing our relationship with the potential partner and ECR represents the distribution for environmental compliance record. The Total Risk output distribution appears in Figure 14.4. As with the qualitative

Total Risk of Partner

FIGURE 14.4 Cumulative frequency plot of total risk for partnership.

model explained in the previous section of this chapter, weights for individual variables or groups of variables also could be applied.

MINI TIME-SERIES MODEL

As outlined in Chapter 12, most projects in a business will have life spans measured in years. Therefore, it is essential that a risk model built to evaluate such an opportunity be capable of assessing not only the present state, but that of years to come as well.

Most real-world time-series models incorporate dozens of variables and, sometimes, hundreds of lines of equations. It is not the point of this section to attempt to impress the reader with intricate models. Rather, it is the aim of the following paragraphs to elucidate just how to approach a time-series risk model. Larger and more complex models adhere to the same principles as are outlined here.

It should be noted that the example presented here is a gross simplification of an actual time-series risk model. In this model, we will consider only two variables: those of Cost and Income. Keep in mind that in a more comprehensive model, the variable Cost would be composed of other variables such as capital expenses (CAPEX), operating expenses (OPEX), and other costs. Likewise, our variable Income would represent the integration of other parameters such as demand, unit price, production rate, and other considerations. Variables such as taxes, discount rate, depreciation, and others typically complicate any real-world analysis.

Our simple model will assess the 5-year project to build a new production facility. Our projected costs, in millions of dollars, for the next 5 years are as follows:

1999	2000	2001	2002	2003
5	3	1.5	0.5	0.5

See Figure 14.5 for a plot of these single values. The $5MM figure for 1999 primarily

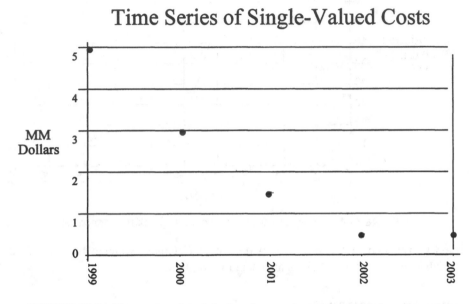

FIGURE 14.5   Time series plot of single values representing projected yearly costs.

represents construction costs (in a real world analysis, construction costs would represent at least one variable). In 2000, we will be operating the plant, but still incurring some capital construction costs. Costs for 2001 are expected to be mainly our operating costs, but the plant is not expected to reach peak operating efficiency until the year 2002. In that year and for subsequent years, our costs are expected to be about $0.5MM per annum.

Our 5-year projection for Income, in millions of dollars, looks like this:

1999	2000	2001	2002	2003
0.1	3	3.5	4	4.5

See Figure 14.6 for a plot of these single values. Income in the first year is expected to be zero or near zero, depending upon the date on which we actually begin production.

The value of 3 for the year 2000 reflects the fact that we expect to be producing goods for most of that year. Gaging demand for similar products, market share, competition, regional income levels (all of which would be separate variables in a comprehensive analysis), etc., we expect our gross sales income to rise by $0.5MM per year.

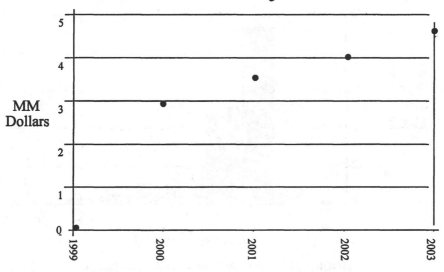

FIGURE 14.6   Time series plot of single values representing projected yearly income.

We have built plants like this one in the past. Our past experience tells us that our first-year costs could be about 10% less than our proposed $5MM value and could exceed the $5MM value by as much as 30% due to labor problems. Therefore, in our risk model, we will, on each iteration (each iteration will treat all 5 years), multiply the $5MM value by a distribution we will call "expcost99" (1999 cost expansion distribution). The expansion distribution will have a minimum value of 0.9, a most likely value of 1.0, and a maximum value of 1.3. We are fairly confident in our yearly estimate for 1999, so the peakedness will be set at 10 (see Figure 14.7). See Chapter 12 for details on some problems associated with time-series analysis, iterations, and expansion factors.

For 2000, we believe that the costs could again be lower than the $3MM figure by 10% but may exceed the $3MM number by as much as 25%. Therefore, we will, on each iteration of the Monte Carlo process, adjust the $3MM value by multiplying it by a distribution named "expcost2000." This distribution will have a minimum value of 0.9, a most likely value of 1.0, and a maximum value of 1.25. We are less confident that the actual value in 2000 will be $3MM, so we set our peakedness to 8 (see Figure 14.8).

The $1.5MM cost projection for the year 2001 is, relative to 1999 and 2000, a fairly firm number because it is based primarily upon operating costs for plants like this one that we have built previously. Costs are expected to be as much as 5% less than the projected $1.5MM value and might exceed that value by as much as 10%. A distribution named "expcost2001" with a minimum of 0.95, a most-likely value

Year 1999 Expansion Distribution

FIGURE 14.7   Year 1999 cost expansion distribution.

Year 2000 Expansion Distribution

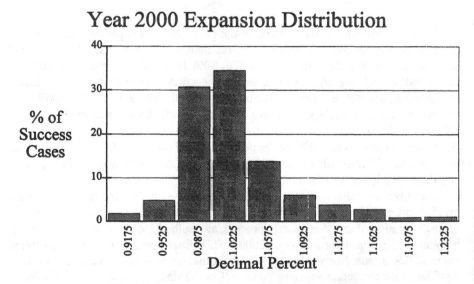

FIGURE 14.8   Year 2000 cost expansion distribution.

of 1.0, and a maximum of 1.1 will be used to convert the $1.5MM coefficient to a distribution (see Figure 14.9). A peakedness of 9 will be used.

The $0.5MM figures for the years 2002 and 2003 are expected to vary by no more than plus or minus 5%. A single expansion distribution called "expcost023" with a minimum of 0.95, a most likely of 1.0, and a maximum of 1.05 will be used

Year 2001 Expansion Distribution

FIGURE 14.9   Year 2001 cost expansion distribution.

to modify the $0.5M values (see Figure 14.10). Peakedness for this distribution was set to 10.

Our projected gross income for 1999 is expected to be $0 or near $0. To expand our $0.1MM value into a distribution, we will multiply the $0.1MM number on each iteration by a truncated distribution. The minimum and most likely values of the truncated distribution will be 0 and the maximum will be 1.01. We named this distribution "expinc99." Peakedness was set to 10 (see Figure 14.11).

For all other years, we expect the income to vary by plus or minus 10%. Therefore, we will create a common expansion distribution called "incexp" with a peakedness of 10, a minimum value of 0.9, and a maximum value of 1.1 (see Figure 14.12).

Total profit for the project would be calculated by the following equations:

$$n = 0; \hspace{4cm} (14.4)$$

$$costexp\{0\} = expcost99;$$

$$costexp\{1\} = expcost00;$$

$$costexp\{2\} = expcost01;$$

$$costexp\{3\} = expcost023;$$

$$costexp\{4\} = expcost023;$$

Years 2002 & 2003 Expansion Distribution

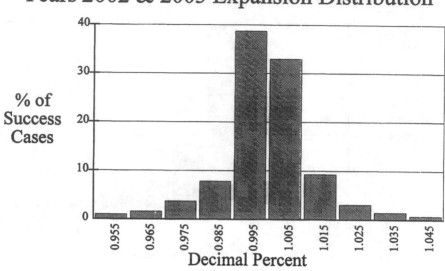

FIGURE 14.10   Cost expansion distributions for the years 2002 and 2003.

Year 1999 Income-Expansion Distribution

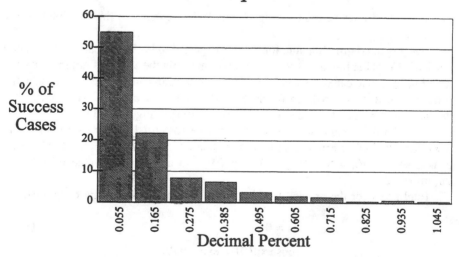

FIGURE 14.11   Year 1999 income-expansion distribution.

```
while n < PERIODS do
  total_cost{n} = cost{n} * costexp{n};
    if n = 0 then
```

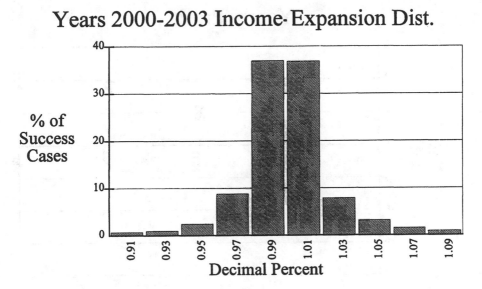

FIGURE 14.12 Income expansion distribution for the years 2000 through 2003.

```
        total_inc{n} = income{n} * expinc99;

    else

        total_inc{n} = income{n} * select(expinc,1);

    endif

profit{n} = total_inc{n} — total_cost{n};

n = n + 1;

enddo
```

The variable PERIODS simply holds the number of periods in this analysis which, in this case, is five. The { } syntax simply allows us to reference a specific period (year, in this case). The "select" function allows us, on each iteration, to sample the "expinc" distribution one time per period. This allows us to have variability in the "expinc" variable across periods, not just across iterations.

The final "profit" plot can be seen in Figure 14.13. As explained in Chapter 12, each of the individual vertical bars can be expanded to display the distribution it represents. If warranted, an NPV (or IRR, or whatever financial measure) function could be called at the end of the program to calculate a net present value for the project.

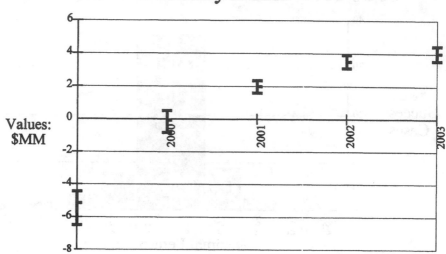

FIGURE 14.13 Time series plot of profit.

FATE-TRANSPORT (PROCESS/HEALTH) MODEL

Fate-transport or process/health models are those that attempt to comprehensively model movements of substances and related exposure considerations. Most fate-transport models attempt to emulate potential human exposure to harmful (or harmful levels of) chemicals or other environmental factors. Typical models and equations attempt to calculate the level of exposure over time. Exposure in humans typically, but not necessarily, falls into several categories. These are

- Ingestion (including soil, groundwater, food, etc.)
- Inhalation
- Dermal (skin) exposure.

The simple example given here is one in which ingestion of a carcinogen will be estimated. The variables in the example will be

- IR = Water ingestion rate (liters/day)
- CW = Chemical concentration of contaminant in water (decimal percentage)
- EF = Frequency of exposure (days/year)
- ED = Duration of exposure (years)
- AT = Averaging time (days)
- BW = Body weight (kilograms)

From these factors the model will calculate

$$\text{INTAKE} = \text{Ingestion of contaminated water (mg/kg-d)} \qquad (14.5)$$

The equation to be used is

$$\text{INTAKE} = (\text{CW} \times \text{IR} \times \text{EF} \times \text{ED})/\text{BW} \times \text{AT}) \qquad (14.6)$$

In the typical deterministic risk assessment, a single value for INTAKE would be calculated from single-valued input parameters. Using the stochastic approach, distributions for IR (Figure 14.14), CW (Figure 14.15), EF (Figure 14.16), ED

FIGURE 14.14   Distribution of water ingestion rate.

(Figure 14.17), AT (Figure 14.18), and BW (Figure 14.19) will be used to calculate an output distribution INTAKE (Figure 14.20).

Uncertainties associated with fate-transport-type models are expressed differently than in typical probabilistic models in which uncertainty is likely to be expressed as a percentage or probability and in which the uncertainty actually is a variable in the probabilistic equations. In a process/health-type model, uncertainties generally are

- Listed as uncertainties with sentences describing the source, magnitude, and effect;
- Related to "gaps" in the data;
- Related to measurement errors;
- Related to sampling errors;
- Related to the use of summary statistics or tabular data;

FIGURE 14.15   Concentration of chemical in water.

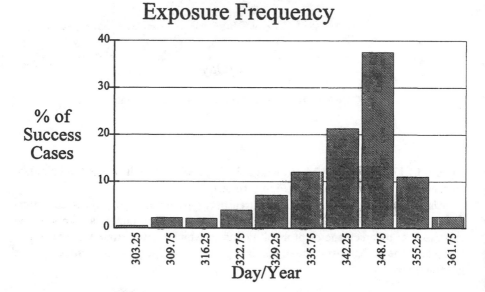

FIGURE 14.16   Distribution of frequency of exposure.

- Related to variability in accepted models;
- Related to conservation in the models.

FIGURE 14.18 Single-valued averaging time.

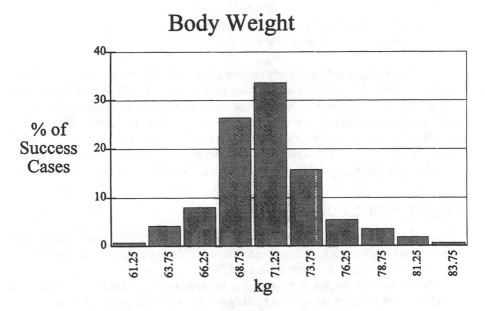

FIGURE 14.19 Distribution of body weight.

- Related to measurement errors;
- Related to sampling errors;
- Related to the use of summary statistics or tabular data;

FIGURE 14.20   Cumulative frequency curve of contaminated water intake.

- Related to variability in accepted models;
- Related to conservation in the models.

Fate-transport models are highly complex affairs and do not generally lend themselves to "sensitivity" analysis as described in this book. Models and equations must be run through several times (changing the coefficient for some parameter each time) to determine the effect of the parameter change on the model result. This is as it should be, given the complexity of the models and the built-in uncertainties. Application of probabilistic techniques to fate-transport-type models can have some beneficial results, but the application is limited due to the nature of most of these models.

DECISION-TREE CONVERSION

Decision trees are popular as a means of generating a deterministic result that incorporates probabilities. Although decision trees do not actually decide anything, they are a very useful decision-making aid.

The visual and graphical nature of a decision tree facilitates discussion of a problem. It forces problem-solving participants to visualize the problem as a more-or-less linear process from a decision point to one or more consequences. This is a significant step in the process of organizing and diagramming a solution to the dilemma.

In addition, the decision tree allows us to take a step into the probabilistic world. This is not so important a point with regard to calculations or resulting answers. It is a great leap forward, however, for people to begin to think in probabilistic terms and to begin to conceptually grapple with the quantification and documentation of uncertainty.

For all its stalwart features, the decision tree is, in the end, a deterministic vehicle. Some software packages have attempted to imbed the decision tree in a stochastic framework. The meld of these two philosophies often results in either a less-than-potent risk assessment package or one that is unacceptably awkward to use.

My philosophy through the years has been to let decision trees be decision trees and risk packages be risk packages. This has stood me in good stead for several reasons. First, the decision tree structure is an excellent means of directing the evolution of a solution to a problem. Its graphical nature is ergonomic and practical. The decision tree also facilitates determination of where in the problem-solving scheme it is important to incorporate probabilities and which, if any, variables should be represented as distributions. Having thus decided, it is an easy and practical practice to convert the decision tree to a full-blown, if you will, probabilistic model.

The first step in the conversion process is, of course, to generate the decision tree. In Figure 14.21 is depicted a very simple decision tree that is intended to help us decide whether to build a plant. In this admittedly oversimplified example, the go/no-go decision on the plant hinges on whether we will have to pay too much for labor.

Contract negotiations with the labor union are in progress. There is a possibility that the negotiations will go swimmingly and we will realize labor costs well within our fiscal limits. However, things may not go so well. We might experience cost increases due to relatively minor changes to the existing contract, or we could realize significant cost increases from a completely new contract.

If costs are prohibitive, we may decide not to build the plant. This decision is not without cost. Payments to partners, reclamation expenses associated with property-ownership transfer, legal costs, and other penalties could make the "don't build" decision, at this point, relatively costly.

In our problem-solving sessions, we have described the decision tree seen in Figure 14.21. We have concluded that all variables on the diagram need to be represented by distributions. The distributions representing total costs including those associated with no labor problems, minor labor problems, and a totally renegotiated contact are shown in Figures 14.22, 14.23, and 14.24, respectively. The projected cost of not building the plant is shown in Figure 14.25.

The simple equation that could be used to convert the decision tree to a Monte Carlo model is

$$\text{buildplant} = (0.6 \times \text{noprob}) + (0.4 \times ((0.2 \times \text{minorprob}) \tag{14.7}$$

$$+ (0.8 \times \text{majorprob})))$$

in which the variables "noprob," "minorprob," and "majorprob" respectively represent the distribution of costs associated with no labor problems, minor labor problems, and totally renegotiated contract. The distribution of costs resulting from this Monte Carlo model is shown in Figure 14.26.

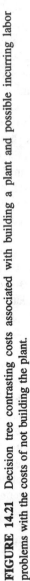

FIGURE 14.21 Decision tree contrasting costs associated with building a plant and possible incurring labor problems with the costs of not building the plant.

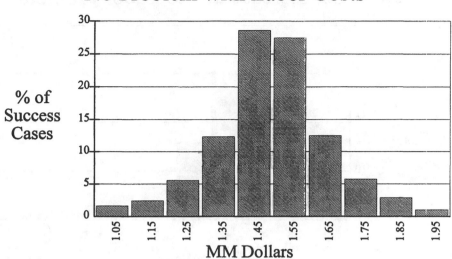

FIGURE 14.22 Distribution of costs associated with no-labor-problem scenario.

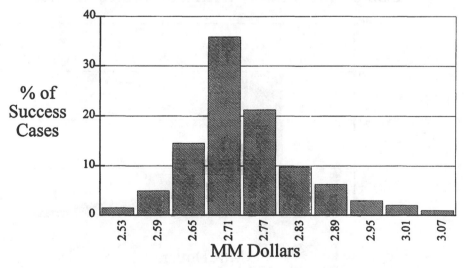

FIGURE 14.23 Distribution of costs associated with minor-labor-problem scenario.

We may be interested in seeing what the Monte Carlo model difference is between building and not building the plant. If so, we might use the simple equation:

$$\text{diff} = \text{buildplant} - \text{dontbuild} \qquad (14.8)$$

Major Cost Increase due to Labor Problem

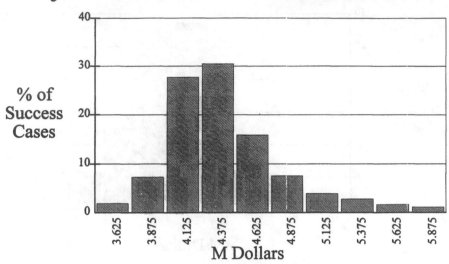

FIGURE 14.24   Distribution of costs associated with a completely new contract.

Costs Associated with Failure to Build

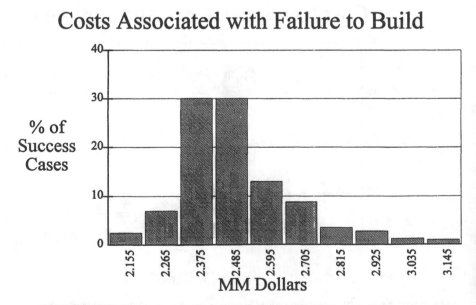

FIGURE 14.25   Distribution of costs associated with failure to build the plant.

This equation will subtract, on each Monte Carlo iteration, the randomly selected cost of not building the plant ("dontbuild" variable) from the randomly selected cost of building the plant. The result of the calculation is shown in Figure 14.27.

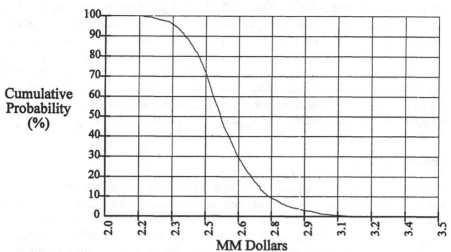

FIGURE 14.26   Cumulative frequency plot of costs associated with building the plant.

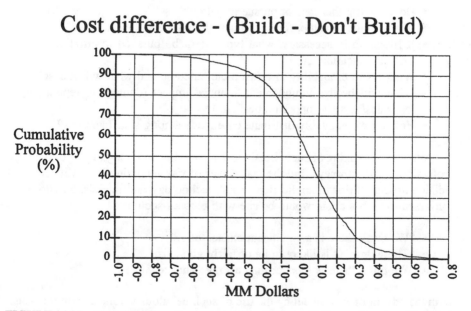

FIGURE 14.27   Cumulative frequency curve of the difference between the costs of building the plant vs. not building the plant.

It should be noted that in this simple example we have used distributions to represent dollar consequences, but have resorted to deterministic estimates of probability. It is a simple matter to represent each probability pair (or group) as dependent distributions. Dependence is necessary because the probabilities need to sum to 1 (or 100, or whatever). So if we were to represent the probability of one branch of the decision tree with a range of 0.2 to 0.6 and on a given iteration the value of 0.35 was selected, then the probability for the complementary branch for that iteration would be 0.65. Note also that this process, in its simplest form, requires that a hierarchy be established for the tree branches. That is, the probability associated with one branch is randomly determined and the other is subject to that randomly determined value. Logic often is added to a model to randomly select, on each Monte Carlo iteration, which branch probability will be drawn from the distribution of probabilities for that branch.

LEGAL MODEL

Risk assessment models are utilized in legal departments and at law firms to address a wide range of problems. The primary purpose for undertaking a legal risk assessment, of course, is to aid attorneys in making decisions. Equally important, however, are the communication, consistency, and documentation aspects of legal risk assessment.

As a legal decision-making tool, risk modeling is a potent supplement to an attorney's knowledge, intuition, and experience. Typical of the types of questions answered are

- Do we settle this case or pursue litigation?
- If we decide to settle, for how much should we settle?
- If litigation is necessary, what type might be pursued (arbitration, jury trial, adjudication, etc.)?
- What is the probability of a favorable outcome and what will it cost?
- What will be the nonjudgment financial impact (damage to reputation, lost sales/business opportunities, etc.)?
- How do we equitably distribute case loads among staff members?

Although the taking of such decisions is a common and integral part of the legal profession, most attorneys find the real power of risk modeling lies in its communication aspects. Prior to application of risk technology and models, an attorney's conversation with a client would be rife with phrases such as

- "We think there is a significant chance of..."
- "The judgment is likely to be of fairly good size..."
- "We think there is a better than average chance..."

With the advent of risk modeling the use of such nebulous verbiage is on the wane. A parameterized legal risk model provides specific items ("variables" if you will) that form the basis of a decision. In addition, each of the items requires qualitative, semi-quantitative, or numerical input from both the attorney and the client. Models that serve as the focus of conversations give the client a better understanding and feeling for the decision-making process. Dialogue also gives the attorney a much clearer view of the client's situation, limits, and preferences. At the end of such conversations, and after running the risk model containing the discussed and agreed-upon inputs, the client can be presented with the entire range of possibilities and the associated probabilities.

Prior to the use of legal risk models, an attorney might advise a client thus: "Mr. Smith, we think that this settlement is likely to cost you about $4 million...." When settlement was finally reached and cost Mr. Smith $5.2 million, Mr. Smith is displeased, to say the least. Using risk-model output, the presettlement conversation with Mr. Smith is more likely to beas follows:

> "Mr. Smith — we put all of the values and judgments that we agreed upon into the risk model. Our uncertainty about several of the input parameters was represented as a distribution of values in each case. You can see from the resulting risk assessment plot that there is a small chance that the settlement could cost as little as $3 million. The most likely outcome is that it will cost around $4 million. However, I would like you to notice that the plot shows that there is a small but significant chance that the settlement could cost as much as $12 million. The $12 million figure can be traced back to the input parameters...."

Now, when the actual cost comes in at $5.2 million, Mr. Smith is not only relieved (rather than perturbed), but understands the circumstances that culminated in the $5.2 million figure.

Sensitivity analysis also is an integral part of legal risk assessment (see Chapter 13 of this book). A sensitivity analysis of a given legal case can indicate to clients which parameters and aspects associated with a litigation are most significant or dire. Such analyses also point out to attorneys which facets of a case deserve the most attention. Much time, effort, and money can be saved by focusing efforts on significant parameters and, equally, by not wasting time tending to items of paltry importance.

Although the aforementioned attributes of legal risk assessment are salient, one of the greatest advantages to the use of risk models in the legal profession is associated with consistency. An initial risk model built for a specific legal purpose will integrate the set of parameters outlined in the model-building discussions. A list of parameters generated by this preliminary dialogue is almost always augmented by experience over time. With repeated application of the model, a comprehensive and concise set of parameters results. Such a model should not be seen as stifling flexibility or creativity or as "cramping someone's style." Each attorney can contribute to the model those attributes that have personal significance. A well-constructed legal model should allow various users to exercise the "required" parts of the model in addition to those parts that represent their personal "style and flair."

A risk model, although flexible, serves as a consistent checklist of considerations so that no significant points are accidentally overlooked. A model also can offer a consistent means of combining input parameters (flexibility can be built in here, too) and of presentation. Department- or firm-wide consistency of understanding and terminology is of great benefit. Models such as that described here can be invaluable when, for example, an attorney new to the firm or department has to be "brought up to speed" concerning the way a particular problem is handled.

Yet another paramount benefit of legal risk assessment is that of documentation and archiving. As with most risk models, it often is most valuable to know *why* a person input the data that they did. A well-constructed risk model should provide for the capture of such information. In addition, the model should offer some means of downloading model input and output data to a database. Such archived information can be invaluable in solving future problems and can be instrumental in foregoing the age-old problems of relearning how to approach a problem and of making the same mistakes over and over again.

The legal risk model presented here is a simple caricature of a "generic" legal model. Any actual model would, of course, be much more detailed and would address a specific legal problem.

In Figure 14.28 is depicted the contributing factor diagram for this fictitious model. The answer variable in the lower right corner of the diagram indicates that we are interested in determining the difference between the cost of settling a case vs. pursuing litigation. Our model is composed of the following simple parameters:

- Probability of a favorable litigation result (expressed as a decimal between 0 and 1). Variable name = LP.
- The internal costs associated with the litigation option. Variable name = LIC.
- The external fees realized from pursuing litigation. Variable name = LEF.
- The settlement amount. Variable name = SA.
- The internal costs associated with settlement. Variable name = SIC.
- The external fees realized from settlement. Variable name = SEC.

Our simple set of equations would be

$$TLC = LP \times (LIC + LEF) \tag{14.9}$$

$$SP = 1 - LP$$

$$TSC = SP \times (SA + SIC + SEC)$$

$$DIFFERENCE = TLC - TSC$$

In the equations, TLC is the total litigation cost, SP is the probability of settlement, TSC is the total settlement cost, and DIFFERENCE is the difference between the cost of litigation and the cost of settlement.

Through conversations with the clients, attorneys have determined that

- Litigation external fees will be a minimum of $200,000, most likely $400,000, and a maximum of $500,000.
- Litigation internal costs will be a minimum of $1,000,000, most likely $3,000,000 and a maximum of $4,500,000.
- Settlement external fees will be a minimum of $50,000, most likely $75,000, and a maximum of $100,000.
- Settlement internal costs will be a minimum of $100,000, most likely $200,000, and a maximum of $300,000.

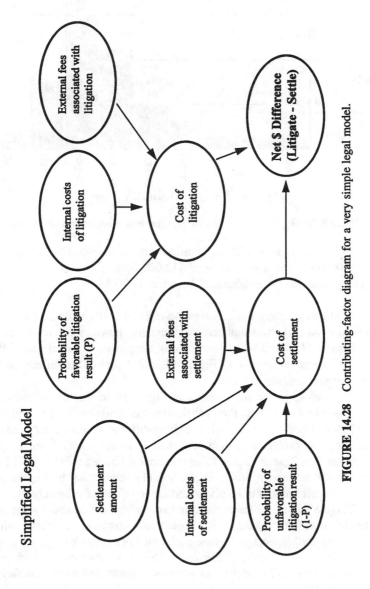

FIGURE 14.28   Contributing-factor diagram for a very simple legal model.

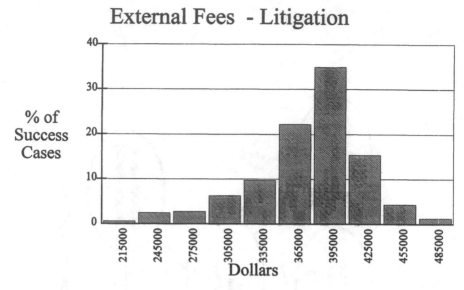

FIGURE 14.29 Frequency plot of external fees associated with litigation.

- Settlement amount will be a minimum of $5,000,000, most likely $10,000,000, and a maximum of $13,000,000.
- The probability of a successful litigation is 75%.

Plots of litigation external fees, litigation internal costs, settlement external fees, settlement internal costs, and settlement amount are shown in Figures 14.29 through 14.33. In Figures 14.34 and 14.35, respectively, are plots of the total cost of litigation and the total cost of settlement. In Figure 14.36 is a plot of the difference between litigation cost and settlement cost.

It can be seen from the information in Figure 14.36 that the probable cost of litigation is nearly the same as the probable cost of settlement. The "Net Amount" shown in Figure 14.36 is the net-risk-weighted-difference cost of our case. The net value subtracts the probable amount less than 0 (in this case) from the probable amount greater than 0 (see Chapter 13 of this book). The net value is $2410.83. This indicates that the settlement option is just slightly better than the litigation option. Given the settlement amounts involved and other values of relatively great magnitude, the $2410.83 value indicates that the two options are essentially the same. To make the decision with regard to settlement and litigation, attorneys might now consider "intangible" aspects of the case such as the personalities of judges, opposing attorneys, and other elements of the case. They also may opt to add additional considerations to the risk model in an attempt to determine more concisely which option is the better.

A sensitivity plot might also be used to resolve the issue. In Figure 14.37 is shown the sensitivity plot (see Chapter 13) for this case. The sensitivity plot indicates that the most significant variable in our analysis is the internal costs associated with litigation. If attorneys can reduce these costs, the analysis certainly would favor

Internal Costs - Litigation

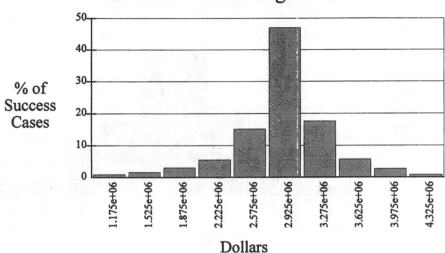

FIGURE 14.30 Frequency plot of internal costs associated with litigation.

External Fees - Settlement

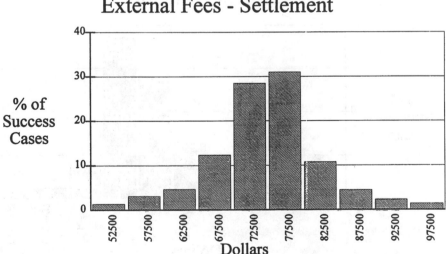

FIGURE 14.31 Frequency plot of external fees associated with settlement.

litigation. Note that the influence of the amount of settlement is secondary to the litigation internal costs. This is in spite of the fact that the settlement costs are of greater absolute magnitude. Note also that spending any time, effort, or money attempting to reduce the costs associated with any other aspects of the case is a waste.

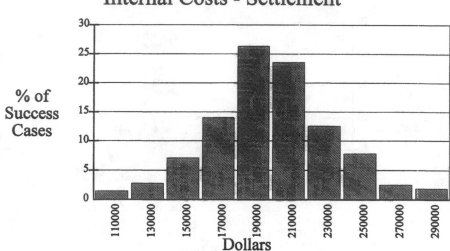

FIGURE 14.32   Frequency plot of internal costs associated with settlement.

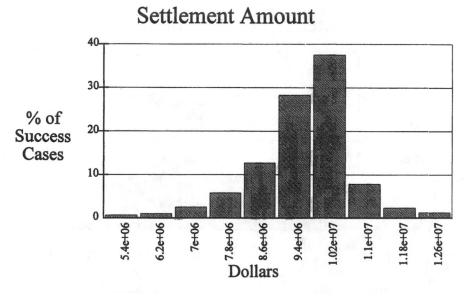

FIGURE 14.33   Frequency plot of settlement amount.

COMPREHENSIVE RISK ASSESSMENT EXAMPLE

Ben is the general manager in charge of construction and operation of the new chemical plant. Construction of the plant is to begin in less than a year. Ben has called together the staff to begin preliminary discussions. Ben and his staff decide

FIGURE 14.34   Cumulative frequency plot of total cost of litigation.

FIGURE 14.35   Cumulative frequency plot of total cost of settlement.

to enlist the help of a risk-process facilitator to aid in planning and organizing this project.

Ben next holds a facilitator-led brainstorming session with his group. The goal of the brainstorming session is to define just what question(s) will be answered by the risk assessment (i.e., just what do they want to know — just what will be

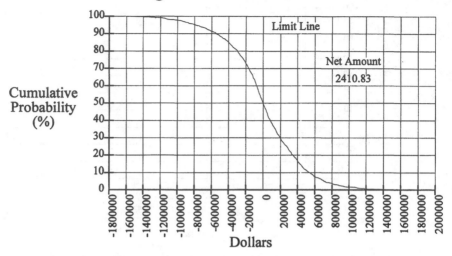

FIGURE 14.36   Cumulative frequency plot of litigation cost minus settlement cost.

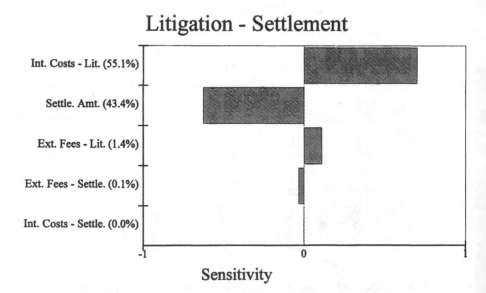

FIGURE 14.37   Sensitivity plot of litigation cost minus settlement cost.

calculated by the risk model?). During the brainstorming session, the facilitator leads the discussion and records all salient ideas put forth. The simple rules for conducting a brainstorming session are

TABLE 14.1

BRAINSTORMING IDEAS

1. What will be the yearly chemical production from the new plant?
2. What are the plans of our competitor chemical producers?
3. We need to convince upper management that we really need the new chemical plant.
4. What will be the cash flow from the new chemical plant?
5. We need to produce our chemical at a competitive price — how will we do that?
6. What will be the NPV and IRR of the new plant project?
7. This project has to be completed quickly enough so that our competition does not build a similar plant first.
8. What are the environmental constraints on chemical-plant production in the new-plant area?
9. How will this project compare with other projects competing for corporate funding?
10. We need to build a plant of sufficient efficiency to maintain our market share — how will we do that?
11. What will be the construction costs for the new plant?
12. We should use in-house engineers to design the plant rather than contract engineers.
13. Will we be able to transport the chemical through all states?

- Gather as many participants/stakeholders as is practical;
- Record all pertinent ideas;
- NO CRITICISM.

At the conclusion of the brainstorming session, the list of proposed issues is as shown in Table 14.1.

Although the risk model may calculate the answers for several of the items in the brainstorming list, the facilitator realizes that the group should focus on just one question to begin with. The risk model can be expanded to include other items after the risk-model foundation is laid for the most important aspect of the study.

To help in the selection of the primary question to be answered by the risk model, the discussion leader decides to use a hierarchy process. To facilitate this process, the discussion leader distributes to each member of Ben's group two sheets of paper. One paper is the list of brainstorming items in Table 14.1. The second paper is divided into two sections: an "NCN" (not considered now) section at the top and a "FOCUS" section at the bottom. Each member of the group is asked to put the number of each statement in Table 14.1 into one of the two sections (NCN or FOCUS items).

Table 14.2 below indicates the results of the voting. As can be seen from the table, each statement received at least one vote. This is typical because the person who offers a statement generally votes for it to be a FOCUS item. There are 20 people in the voting group. As a rule of thumb, the facilitator generally considers a statement to be a serious FOCUS item if it receives half or more of the votes. In this case, the statements that received 10 or more FOCUS votes are

TABLE 14.2
Not Considered Now and Focus-Item Votes

Statement #	NCN # votes	FOCUS # votes
1	19	1
2	12	8
3	10	10
4	9	11
5	19	1
6	7	13
7	19	1
8	17	3
9	8	12
10	13	7
11	19	1
12	19	1
13	17	3

#3 We need to convince upper management that we really need the new chemical plant.
#4 What will be the cash flow from the new chemical plant?
#6 What will be the NPV and IRR of the new plant project?
#9 How will this project compare with other projects competing for corporate funding?

The facilitator lists these four statements on a flip chart and leads the group through a discussion of the items. In the discussion, the group concludes that statement #4 is "covered" by statement #6 because NPV and IRR are calculations that consider cash flow. After much discussion and debate, it also is decided that what will convince upper management that the plant is needed (statement #3) is the fact that we can produce and sell economic quantities of our chemical. This will be reflected in NPV and IRR values, and so proponents of statement #3 agree that favorable NPV and IR values resulting from the risk model will suffice. Similarly, proponents of statement #9, after much discussion, agree that management likely will use NPV and IRR as their main project-ranking tool and, therefore, the calculation of NPV and IRR likely will satisfy their concern about competing corporate projects.

The facilitator realizes that he is lucky this time. In past sessions with other groups, much more contentious discussion ensued that prompted several rounds of voting. In such sessions it has been necessary to have the group vote a second time on only the statements that were selected as FOCUS items in the first voting round (statements 3, 4, 6, and 9 in the current session). In some situations, even after several rounds of voting, more than one statement persisted in the FOCUS group. In such situations, the facilitator has had to have the group decide whether they truly had multiple problems to solve and whether more than one risk model needed to be built to solve them.

FIGURE 14.38   Initial contributing-factor diagram.

In this case, using the hierarchy method, the group settles on the question "What will be the NPV and IRR of the new-plant project?" as the first question they will attempt to answer with the risk assessment model. To begin to build the model, the facilitator begins to help the group build a contributing-factor diagram. A typical process for building a contributing factor diagram generally posts the "answer variable" first. Items that lead to the calculation of (items that contribute to) the answer variable are then posted. Figures 14.38 and 14.39 show initial contributing-factor diagrams (see Chapter 6 for detailed development of these diagrams).

The group built these diagrams keeping in mind

- The answer variable units are dollars;
- All other variables posted will be elements of a system of equations that will calculate, eventually, dollars.

In Figure 14.38 only the answer variable and two major contributing factors are posted. Each of the two contributing factors in Figure 14.38 are discussed by the group and expanded into the diagram shown in Figure 14.39.

Although a much more detailed contributing-factor diagram likely could be built, the relatively simple contributing-factor diagram shown in Figure 14.39 is the one Ben will use for the sake of this example. The simplified contributing-factor diagram

FIGURE 14.39 Completed contributing-factor diagram.

will be used because it is not practical to present an exhaustive treatment of the problem in this format.

Note that in the contributing-factor diagram, units are given for each factor. Note also that some units are, for example, X/Day while others are X/Year. When listing units, always consider what units are most convenient for the user. For example, when asking how many miles a person puts on their car, it is most likely that they will respond with a value like "12,000 per year." However, when asking someone how much they spend for lunch, they are more likely to indicate "$6.00 per day."

In the same vein, while some variables are broken down into component parts on the contributing-factor diagram, the component parts may not be presented in the risk model. For example, in the contributing-factor diagram, "Utility Costs" is composed of "Kilowatt Hours per Year" and "Cents per Kilowatt Hour." It will be decided in our model that it is easier for the user to estimate the total utility costs per year rather than to guess at kilowatt-hour values.

In addition, it may be decided that we should split-out some of the factors into a separate model. For example, we may decide to make a separate "Construction Costs" risk model, the answers from which would be included in our more comprehensive model.

Ben knows that he has to make preliminary calculations of pertinent costs/incomes. The list of major risk-model variables derived from the contributing-factor diagram is

- Construction costs
- Hourly wage
- Number of workers
- Utility costs
- Price per ton
- Tons per day
- Discount rate
- Hours per day
- Days per year
- Corporate tax rate

The team decides that because construction of the plant is to begin soon, the variable "Construction Costs" should be set up as a separate model. In initial meetings with the construction-cost-estimating group, Ben's team and the construction team engage in a "ranging" exercise for construction-cost components. For each component, minimum, most likely, maximum, and peakedness (confidence in the most-likely value) value, are generated. These are as shown in Table 14.3.

The team determines that the peakedness values should be relatively low, indicating their relatively low confidence that the most-likely values they had listed would, in the end, actually turn out to be the actual costs. The team was about to enter the totals from Table 14.3 into a spreadsheet/risk model when Jerry, Ben's risk expert, indicated that the summed construction costs should not be entered into the risk model.

TABLE 14.3
Construction-Cost Distributions

Component	Min.(MM$)	M.L.(MM$)	Max.(MM$)	Peakedness
Pipe	8.1	9.0	9.9	2
Steel	0.9	1.0	1.1	1
Compressors	10.8	12.0	13.2	4
Concrete	2.25	2.5	2.75	3
Exchangers	2.25	2.5	2.75	2
Labor	22.5	25	27.5	1
TOTALS	46.8	52	57.2	

Pipe Costs

FIGURE 14.40   Pipe-cost distribution used as input for construction-cost model.

Jerry points out that if the individual construction costs are to be entered into the risk model, they certainly could be entered as individual ranges as shown in Table 14.3. However, because they have decided that construction costs are to be summed outside the main risk model, they must be probabilistically summed in a separate risk model. In their risk-modeling system, they build distributions for the construction-cost components shown in Table 14.3 (Figures 14.40 through 14.45) which are summed by the Monte Carlo process using the simple equation

$$\text{total\_const\_cost} = \text{pipe} + \text{steel} + \text{compressors} + \text{concrete} + \text{exchangers} \quad (14.10)$$

The output distribution for total_const_cost is shown in Figure 14.46. This distribution will later be passed as an input distribution to the comprehensive risk model.

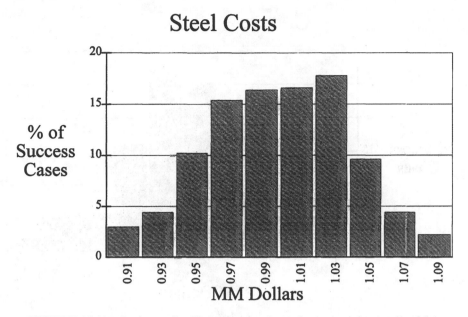

FIGURE 14.41   Steel-cost distribution used as input for construction-cost model.

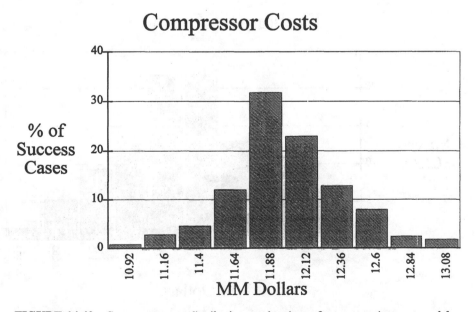

FIGURE 14.42   Compressor-cost distribution used as input for construction-cost model.

FIGURE 14.43   Concrete-cost distribution used as input for construction-cost model.

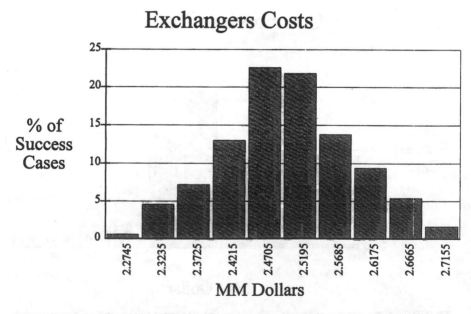

FIGURE 14.44   Exchangers-cost distribution used as input for construction-cost model.

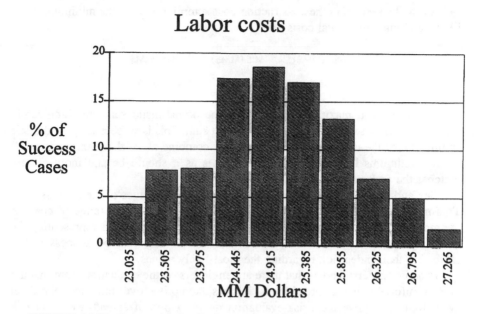

FIGURE 14.45   Labor-cost distribution used as input for construction-cost model.

FIGURE 14.46   Total-construction-cost curve.

As can be seen from the construction-cost-summing model, the minimum, most likely, and maximum total costs are

Min.(MM$)	M.L.(MM$)	Max.(MM$)
48.8	52.0	55.4

The minimum and maximum values from the probabilistic sum are significantly different from those resulting from the straight sum. This is so because probabilistic summation of the distributions will not likely combine all of the maximums or all of the minimums in a single iteration. This is as it should be and more closely matches the real-world situation.

Ben and Jerry understand that they are in the initial stages of their assessment. This means that they don't have a very good idea regarding the range of costs or the most likely costs. Therefore, Jerry expects that the distributions representing the various parameters will not only be broad, but will have low peakedness values indicating their indecision regarding the most-likely values.

In addition, Jerry knows that there are engineers on the evaluation team that are more comfortable using spreadsheets. Jerry is attempting, over time, to wean them away from the spreadsheet-risk-evaluation mode (a mix of spreadsheets and risk models). He decides, for the time being, to generate all models in spreadsheet/risk-model mode. The risk-model-only and the risk model/spreadsheet/risk-model approaches will generate the same results, but will have drastically different run times and user interaction. The risk-model-alone approach is much more efficient and faster to execute.

In the model, First-Year Construction Costs (FYCC) will be read in as a distribution imported from the constriction-cost risk model. The variables Hourly Wage (HW), Number of Workers (NW), Utility Costs (UC), Price Per Ton (PPT), Tons Per Day (TPD), Construction-Cost Multiplier (CCM), and Percent of Labor force that is Construction labor (PLC) will be represented as time-series variables. Discount Rate (DR), Hours Per Day (HPD), Days Per Year (DPY), and Corporate Tax Rate (CTR) will be represented as either single values or single-period distributions. The values for these variables will be as shown in Table 14.4. The year-by-year initial values for the time-series variables are shown in Table 14.5. These values have been entered into our spreadsheet and our spreadsheet-driving risk program.

TABLE 14.4
Single-Valued or Single-Period Variables

	Min.	M.L.	Max.
DR (Integer %)		11	
CTR (Decimal %)		0.5	
HPD (Hours)	6.5	7.0	7.2
DPY (Days)	198	200	202

TABLE 14.5
Time-Series Variables

YEARS	1998	1999	2000	2001	2002	2003	2004	2005	2006	2007
FYCC (MM$)	Distribution imported from construction-cost risk model									
HW($)	12	12.5	13	13.5	14	14.5	15	15.5	16	16.5
NW (#)	65	45	30	25	20	20	20	20	20	20
UC (MM$)	.5	.75	1	1	1	1	1	1	1	1
PPT (MM$)	.0002	.00021	.00022	.00023	.00024	.00025	.00026	.00027	.00028	.00029
TPD (Tons)	30	500	550	600	600	600	600	600	600	600
PLC (%)	0.9	0.3	0.05	0.05	0.05	0.05	0.05	0.05	0.05	0.05
CCM (%)	1.0	0.3	0.1	0.1	0.004	0.004	0.004	0.004	0.004	0.004

Because the initial stage of the project is primarily a construction phase, the manager whose OK is required to proceed with the project is an engineer. Although known to be a stickler for technical details, she is not known to be so demanding with regard to beginning-stage costs (because we are so early in the planning stage of the project). Most single values in Table 14.5 need to be expressed as distributions. Because we are early in the project and because of the manager's preferences, we will expand each time-series single value for a variable by using a single expansion distribution. The variable-by-variable expansion distributions for the variables found in Table 14.5 are shown in Table 14.6.

The distributions built from the values in Table 14.6 will be used as multipliers for the corresponding single values in each year. This will expand, for example, the 1998 tons-per-day value of 10 from a minimum of 9 to a maximum of 12.

For the sake of simplicity, in this example, we will

- Not apply taxes to "negative net income" (usually done in an attempt to offset taxes elsewhere in the corporation);
- Not consider depreciation.

The spreadsheet used for this model is shown in Figure 14.47. Note that the data in the spreadsheet are not exactly the same as those listed in Tables 14.5 and 14.6.

TABLE 14.6
Expansion Distributions for Time-Series Variables

	Min.	M.L.	Max.	Peakedness
HWexp	0.9		1.1	5
NWexp	0.95		1.05	5
UCexp	0.95		1.05	5
PPTexp	0.95		1.05	5
TPDexp	0.9	1.0	1.2	5

A	B	C	D	E	F	G	H	I	J	K	L
YEARS		1998	1999	2000	2001	2002	2003	2004	2005	2006	2007
Construction Costs (MM$)		52	15.6	5.2	5.2	0	0	0	0	0	0
Hourly Wage ($/hr)		12	12.5	13	13.5	14	14.5	15	15.5	16	16.5
Number of Workers (Integer)		65	45	30	25	20	20	20	20	20	20
Utility Costs (MM$)		0.5	0.75	1	1	1	1	1	1	1	1
Price per Ton (MM$)		0.0002	0.00021	0.00022	0.00023	0.00024	0.00025	0.00026	0.00027	0.00028	0.00029
Tons per Day (Tons)		10	500	550	600	600	600	600	600	600	600
Construction Cost Mult. (%)		1	0.3	0.1	0.1	0	0	0	0	0	0
Percent Labor Const. (%)		0.9	0.3	0.05	0.05	0.05	0.05	0.05	0.05	0.05	0.05
Revenue (MM$)		0.4	21	24.2	27.6	28.8	30	31.2	32.4	33.6	34.8
Expense (MM$)		0.61232	1.067	1.53352	1.4617	1.38304	1.39672	1.4104	1.42408	1.43776	1.45144
Expenditures (MM$)		53.01088	15.843	5.22808	5.2243	0.02016	0.02088	0.0216	0.02232	0.02304	0.02376
Discount Rate (Decimal %)	0.11										
#Hours per Day (Floating)	7.2										
#Days per Year (Integer)	200										
Corp. Tax Rate (Decimal %)	0.5										
First-Year Const. Cost (MM$)	52										
Net Income Before Tax (MM$)		-0.21232	19.933	22.66648	26.1383	27.41696	28.60328	29.7896	30.97592	32.16224	33.34856
After-Tax Cash ($)		-53.2232	-5.8765	6.10516	7.84485	13.68832	14.28076	14.8732	15.46564	16.05808	16.65052
NPV	-1.311904										
IRR	10%										

FIGURE 14.47　Initial spreadsheet to be driven by the risk model.

In spreadsheets that are driven by risk models, data are overridden on each iteration. Therefore, it is important that the data in the tables be entered only in the risk model. The equations used in the spreadsheet-driving risk model are

$$c:\text{\textbackslash directory1\textbackslash directory2\textbackslash example.xls;} \qquad (14.11)$$

```
n = 0;
while n < PERIODS do
CC{n} = FYCC * CCM{n};
HW{n} = HW{n} * select(HWexp,1);
NW{n} = NW{n} * select(NWexp,1);
UC{n} = UC{n} * select(UCexp,1);
PPT{n} = PPT{n} * select(PPTexp,1);
TPD{n} = TPD{n} * select(TPDexp,1);
n = n + 1;
enddo
paramin FYCC B18;
paramin CCM C9:19;
paramin PLC C10:L10;
paramin HW C4:L4;
paramin NW C5:L5;
paramin UC C6:L6;
paramin PPT C7:L7;
paramin TPD C8:L8;
paramin DR B14;
paramin HPD B15;
paramin DPY B16;
paramin CTR B17;
paramout netincome C19:L19;
paramout atc C21:L21;
paramout npv B23;
paramout irr B24;
run;
irr = (irr * 100) — (dr * 100);
```

where

- c:\directory1\directory2\example.xls; is the path to the Excel spreadsheet.
- PERIODS is the number of periods (years) over which we will consider the plant.

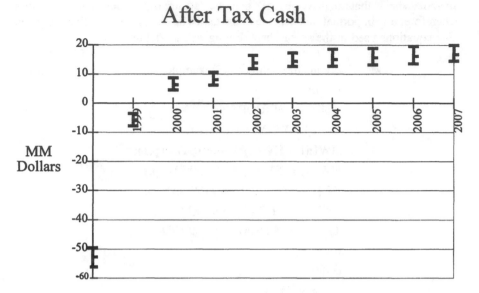

FIGURE 14.48   Time-series plot of after-tax cash fron initial risk model.

- {n} is the period (year) index.
- select (x,1) is a function that allows resampling of distributions across time periods, not just for each iteration.
- paramin and paramout are commands that relate risk-model distributions to spreadsheet cells.

After-tax cash flow is shown in Figure 14.48. The resultant distributions for IRR (minus the discount rate) and NPV are shown in Figures 14.49 and 14.50, respectively. Note that the net IRR (minus the discount rate) is just barely positive. The net NPV value also is of small magnitude.

Because the IRR value for the project is positive, the NPV net risk-weighted value is positive (although, just barely positive), and the probability of obtaining a positive NPV is around 50, the engineer manager (who has go/no-go authority) allows us to continue with the project. She does this, however, with the warning that the next manager with whom we will have to interact is an economist and, therefore, is likely to have a dim view of a project that has about a 50% chance of a positive NPV.

During the time it took Ben and Jerry to complete their initial analysis of the project (about a 6-month period for this project), a lot of things have happened. First, an environmental group has objected to construction of the proposed facility. The environmental group has some misguided ideas about the danger to the community from our proposed plant. The group can't stop the effort completely, but they can file papers in court to delay the project. They have, however, agreed to sit down with the company to discuss the realistic expectations with regard to hazards and what the company might do to mitigate these things.

Internal Rate of Return

FIGURE 14.49 Plot of internal-rate-of-return distribution from initial risk model.

Net Present Value

FIGURE 14.50 Plot of net present value from initial risk model.

The new parameters that this development will add to the risk model(s) are Probability of Delay (PD), Delay Time (DT), Dollars Per Day Delayed (DPDD), and Additional Construction Costs (ACC) associated with mitigation of perceived and real environmental problems (see Table 14.7).

TABLE 14.7
Distributions for New Parameters

			Min.		M.L.		Max.		Peakedness	
PD (decimal %)			0.1		0.15		0.3		10	
DT (Days)			30		60		90		5	
DPDD ($)			10000		15000		20000		9	

	1998	1999	2000	2001	2002	2003	2004	2005	2006	2007
ACC (MM$)	2	1	0.5	0	0	0	0	0	0	0

In addition, during the 6-month period, the political problem of zoning has reared its ugly head. The proposed plant site is not too distant from a residential area (thus, the environmental objections). Completely unrelated to the project, the local housing developers are pushing the legislative bodies to include the potential plant site in the residential zone. If the plant site is zoned as residential, the project will be killed. The company is not without influence with the local government. Many jobs will be created if the new plant is built. Millions of dollars per year will be pumped into the local economy by the new facility. Therefore, the company believes that there is only about a 15% chance that the project will be killed altogether. The variable Ben will use to represent the probability of failure due to zoning is PZF.

Corporate intelligence indicates that the competition may be planning to build a plant similar to Ben's to capture the market share that Ben needs for this to be an economic venture. It has been decided that if the competition begins to build a plant prior to our major construction-spending phase (i.e., while Ben is still in the planning phase), then he will not build the plant. The best estimate from the corporation regarding the competition's plant is that there is about a 10% chance that they will beat Ben to the punch. Therefore, there is a 10% chance of abject project failure due to the competition. Ben will use the variable FDTC to represent failure due to competition.

Also during the 6-month period, new technology has been developed that can significantly boost production. The cost of the new technology supplanting older technology will not be significant (construction costs for the new equipment and construction costs for the old equipment are about the same). The projected New Technology Increased Production (NTIP) figures are shown in Table 14.8.

TABLE 14.8
Time-Series for Increased Production from New Technology

	1998	1999	2000	2001	2002	2003	2004	2005	2006	2007
NTIP (T/D)	0	50	75	85	85	85	85	85	85	85

TABLE 14.9
Expansion Distributions for Prices

	Min.	M.L.	Max.	Peakedness
PPTexp 1998	0.9	1.0	1.1	9
PPTexp 1999	0.85	1.0	1.3	7
PPTexp 2000	0.75	1.0	1.4	5
PPTexp 2001–2007	0.6	1.0	1.4	0

As mentioned previously, the new manager is an economist. This person thinks that he has a better grip on how prices will change from year to year. The manager and a project economist have suggested the expansion distributions for prices shown in Table 14.9.

The new expansion distributions are shown in Figures 14.51 through 14.54. Note the "tighter" range in the distributions and the generally higher peakedness values (except for the 2001–2007 time span). The restricted range and generally higher peakedness values reflect the economist's confidence in their ability to predict prices in the coming years.

The risk model must be modified to reflect

- Probability of environmental delay
- Delay time
- Dollars per day of delay
- Additional construction costs to mitigate environmental problems

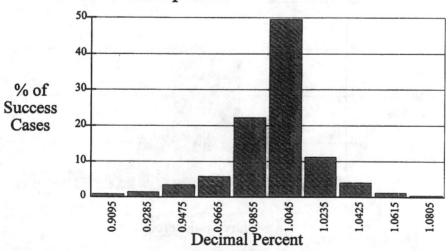

FIGURE 14.51 Expansion distribution for 1998 prices. The deterministic time-series value is multiplied by this distribution to expand the time-series value into a distribution.

FIGURE 14.52 Expansion distribution for 1999 prices. The deterministic time-series value is multiplied by this distribution to expand the time-series value into a distribution.

FIGURE 14.53 Expansion distribution for 2000 prices. The deterministic time-series value is multiplied by this distribution to expand the time-series value into a distribution.

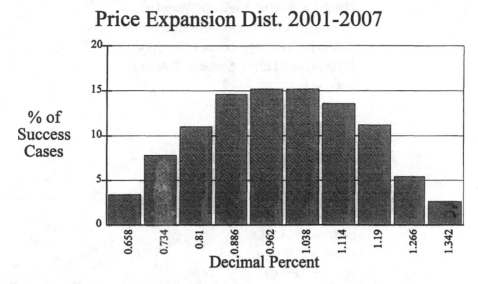

FIGURE 14.54 Expansion distribution for 2001–2007 prices. The deterministic time-series value is multiplied by this distribution to expand the time-series value into a distribution.

- Probability of abject project failure due to zoning
- Probability of abject project failure due to the competition
- New-technology-induced increased production
- New price-expansion distributions

The new set of equations used in the risk model are as follows:

```
c:\directory1\directory2\example2.xls;                          (14.2)
n = 0;
while n < PERIODS do
if n = 0 then
  PPT{n} = PPT{n} * PPTexp98;
elseif n = 1 then
  PPT{n} = PPT{n} * PPTexp99;
elseif n = 2 then
  PPT{n} = PPT{n} * PPTexp00;
else
  PPT{n} = PPT{n} * select(PPTexp0107,1);
endif
CC{n} = FYCC * CCM{n};
HW{n} = HW{n} * select(HWexp,1);
```

```
NW{n} = NW{n} * select(NWexp,1);
UC{n} = UC{n} * select(UCexp,1);
TPD{n} = TPD{n} * select(TPDexp,1);
NTIP{n} = NTIP{n} * select(NTIPexp,1);
n = n + 1;
enddo

paramin FYCC B18;
paramin CCM C9:19;
paramin PLC C10:L10;
paramin HW C4:L4;
paramin NW C5:L5;
paramin UC C6:L6;
paramin PPT C7:L7;
paramin TPD C8:L8;
paramin DR B14;
paramin HPD B15;
paramin DPY B16;
paramin CTR B17;
paramin PD B20;
paramin DT B21;
paramin DPDD B22;
paramin ACC C23:L23;
paramin NTIP C24:L24;
paramout netincome C27:L27;
paramout atc C29:L29;
paramout npv B31;
paramout irr B32;
run;
irr = (irr * 100) — (dr * 100);
```

where

- c:\directory1\directory2\example2.xls; is the path to the Excel spread-sheet.
- PERIODS is the number of periods (years) over which we will consider the plant.
- {n} is the period (year) index.

- select $(x,1)$ is a function that allows resampling of distributions across time periods, not just for each iteration.
- paramin and paramout are commands that relate risk-model distributions to spreadsheet cells.

The spreadsheet being "driven" by the risk model appears in Figure 14.55. The output NPV and IRR plots are shown in Figures 14.56 and 14.57, respectively.

In Figure 14.56, it can be seen that the NPV curve still extends slightly into the negative-dollar range. The extent of the NPV curve into the negative range is much reduced relative to the NPV curve produced by the first model. Although prices have not changed significantly, the increase in production due to the introduction of new technology has had a relatively great positive impact on NPV. The net risk-weighted NPV value reflects this. Note also the Y-axis intercept for the NPV cumulative frequency curve. The chance of failure from zoning is 15% (0.15). The chance of failure due to the competition building first is 10% (0.1). The total chance of success is thus:

$$TCS = (1 - 0.15) \times (1 - 0.1) = 0.85 \times 0.9 = 0.765 \qquad (14.13)$$

This is, of course, a 76.5% chance of success. The net risk-weighted value considers the 23.5% chance of abject failure in its calculation. This same impact can be seen in the IRR cumulative frequency curve. The significantly positive NPV and IRR net risk-weighted values (in spite of the 23.5% chance of abject failure) indicate that the project likely should be pursued.

A	B	C	D	E	F	G	H	I	J	K	L
YEARS		1998	1999	2000	2001	2002	2003	2004	2005	2006	2007
Construction Costs (MM$)		52	15.6	5.2	5.2	0	0	0	0	0	0
Hourly Wage ($/hr)		12	12.5	13	13.5	14	14.5	15	15.5	16	16.5
Number of Workers (Integer)		65	45	30	25	20	20	20	20	20	20
Utility Costs (MM$)		0.5	0.75	1	1	1	1	1	1	1	1
Price per Ton (MM$)		0.0002	0.00021	0.00022	0.00023	0.00024	0.00025	0.00026	0.00027	0.00028	0.00029
Tons per Day (Tons)		10	500	550	600	600	600	600	600	600	600
Construction Cost Mult. (%)		1	0.3	0.1	0.1	0	0	0	0	0	0
Percent Labor Const. (%)		0.9	0.3	0.05	0.05	0.05	0.05	0.05	0.05	0.05	0.05
Revenue (MM$)		0.4	23.1	27.5	31.74	33.12	34.5	35.88	37.26	38.64	40.02
Expense (MM$)		0.61232	1.067	1.53352	1.4617	1.38304	1.38672	1.4104	1.42408	1.43776	1.45144
Expenditures (MM$)		55.14588	16.978	5.86308	5.3593	0.15516	0.15588	0.1566	0.15732	0.15804	0.15876
Discount Rate (Decimal %)	0.11										
#Hours per Day (Floating)	7.2										
#Days per Year (Integer)	200										
Corp. Tax Rate (Decimal %)	0.5										
First-Year Const. Cost (MM$)	52										
Prob. of Delay (Decimal %)	0.15										
Delay Time - Env. Prob. (Days)	60										
$/Day of Delay ($)	15000										
Additional Const. Costs (MM$)		2	1	0.5	0	0	0	0	0	0	0
New Tech. Inc. Prod. (Tons/Day)		0	50	75	90	90	90	90	90	90	90
Net Income Before Tax (MM$)		-0.21232	22.033	25.96648	30.2783	31.73696	33.10328	34.4696	35.83592	37.20224	38.56856
After-Tax Cash ($)		-55.3582	-5.9615	7.12016	9.77985	15.71332	16.39576	17.0782	17.76064	18.44308	19.12552
NPV	4.9069229										
IRR	13%										

FIGURE 14.55   Final spreadsheet to be driven by the risk assessment model.

Net Present Value

FIGURE 14.56   Final net-present-value distribution.

Internal Rate of Return

FIGURE 14.57   Final internal-rate-of-return distribution.

Selected Readings

INTRODUCTION

Skinner, David C., *Introduction to Decision Analysis — Beginning Coursebook*, Probabilistic, 1996.

CHAPTER 1

Finkler, S. A., *Finance and Accounting for Nonfinancial Managers*, Prentice Hall, Englewood Cliffs, NJ, 1992.

Nijhuis, H. J., and Baak, A. B., A Calibrated Prospect Appraisal System, *Proceedings of Indonesian Petroleum Association* (IPA 90-222), Nineteenth Annual Convention, 69-83, October, 1990.

Rodricks, J. V., *Calculated Risks — Understanding the Toxicology and Human Health Risks of Chemicals in our Environment*, Cambridge University Press, Cambridge, 1994.

CHAPTER 2

Atkins, S., *The Name of Your Game — Four Game Plans for Success at Home and Work*, Ellis & Stewart, Beverly Hills, CA, 1993.

Jasanoff, S., Bridging the Two Cultures of Risk Analysis, *Risk Analysis* 2, 123-129, 1993.

Nichols, N., Scientific Management at Merk: An Interview with CFO Judy Lewent, *Risk Analysis* 1, 89-99, 1994.

Pasmore, W. A., *Creating Strategic Change — Designing the Flexible, High-Performing Organization*, John Wiley & Sons, New York, 1994.

Wheatley, M. J., *Leadership and the New Science — Learning about Organization from an Orderly Universe*, Berrett-Koehler, San Francisco, 1993.

CHAPTER 4

Atkins, S., *The Name of Your Game — Four Game Plans for Success at Home and Work*, Ellis & Stewart, Beverly Hills, CA, 1993.

Block, P., *The Empowered Manager: Positive Political Skills at Work*, Jossey-Bass, San Francisco, 1990

Jasanoff, S., Bridging the Two Cultures of Risk Analysis, *Risk Analysis* 2, 123-129, 1993.

Johnson, B., *Polarity Management — Identifying and Managing Unsolvable Problems*, HRD Press, Amherst, MA, 1992.

Maister, D. H., *Managing the Professional Service Firm*, The Free Press, New York, 1993.

Morgan, G. M., and Henrion, M., *Uncertainty — A Guide to Dealing with Uncertainty in Quantitative Risk and Policy Analysis*, Cambridge University Press, Cambridge, 1992.

Pasmore, W. A., *Creating Strategic Change — Designing the Flexible, High-Performing Organization,* John Wiley & Sons, New York, 1994.

Roussel, P. A., Saad, K. N., and Erickson, T. J., *Third Generation R&D — Managing the Link to Corporate Strategy,* Harvard Business School Press, Boston, 1991.

CHAPTER 5

Block, P., *The Empowered Manager: Positive Political Skills at Work,* Jossey-Bass, San Francisco, 1990.

Johnson, B., *Polarity Management — Identifying and Managing Unsolvable Problems,* HRD Press, Amherst, MA, 1992.

Maister, D. H., *Managing the Professional Service Firm,* The Free Press, New York, 1993.

Morgan, G. M., and Henrion, M., *Uncertainty — A Guide to Dealing with Uncertainty in Quantitative Risk and Policy Analysis,* Cambridge University Press, Cambridge, 1992.

Pascale, R. T., *Managing on the Edge — How the Smartest Companies Use Conflict to Stay Ahead,* Simon & Schuster, New York, 1990.

Skinner, D. C., *Introduction to Decision Analysis — Beginning Coursebook,* Probabilistic, 1996.

CHAPTER 6

Skinner, D. C., *Introduction to Decision Analysis — Beginning Coursebook,* Probabilistic, 1996.

CHAPTER 7

Pascale, R. T., *Managing on the Edge — How the Smartest Companies Use Conflict to Stay Ahead,* Simon & Schuster, New York, 1990.

Pasmore, W. A., *Creating Strategic Change — Designing the Flexible, High-Performing Organization,* John Wiley & Sons, New York, 1994.

Senge, P. M., *The Fifth Discipline — The Art and Practice of the Learning Organization,* Doubleday Currency, New York, 1990.

Wheatley, M. J., *Leadership and the New Science — Learning about Organization from an Orderly Universe,* Berrett-Koehler, San Francisco, 1993.

CHAPTER 8

Bayes, T., An Essay Towards Solving Problem in the Doctrine of Chances, *Philosophical Transactions of Royal Society* 53, 1763.

Carbonell, J. G., Editor, *Machine Learning — Paradigms and Methods,* The MIT Press, Cambridge, MA, 1990.

Davis, J. C., *Statistics and Data Analysis in Geology,* Volume 2, John Wiley & Sons, New York, 1986.

Harbaugh, J. W., Davis, J. C., and Wendebourg, J., *Computing Risk for Oil Prospects: Principles and Programs,* Pergamn Press, Tarrytown, NY, 1995.

Newendorp, P. D., *Decision Analysis for Petroleum Exploration,* Pennwell, Tulsa, OK, 1975.

Nilsson, N. J., *Principles of Artificial Intelligence,* Tioga, Palo Alto, CA, 1980.

Schmucker, K. J., *Fuzzy Sets, Natural Language Computations, and Risk Analysis,* Computer Science Press, Rockville, MD, 1994.

CHAPTER 9

Bernstein, P. L., *Against the Gods — The Remarkable Story of Risk,* John Wiley & Sons, New York, 1996.

Dore, A. G., and Sinding-Larsen, R., Editors, *Quantification and Prediction of Hydrocarbon Resources,* from the Norwegian Petroleum Society (NPF), Special Publication No. 6, Elsevier, Amsterdam, 1996.

Meyer, H. A., Editor, *Symposium on Monte Carlo Methods,* John Wiley & Sons, New York, 1954.

CHAPTER 10

Davis, J. C., *Statistics and Data Analysis in Geology,* Volume 2, John Wiley & Sons, New York, 1986.

Ebdon, D., *Statistics in Geography — A Practical Approach,* Basil Blackwell, Oxford, 1981.

Harbaugh, J. W., Davis, J. C., and Wendebourg, J., *Computing Risk for Oil Prospects: Principles and Programs,* Pergamn Press, Tarrytown, New York, 1995.

Hayslett, H. R., *Statistics Made Simple,* Doubleday, New York, 1968.

Meyer, M. A., and Booker, J. M., *Eliciting and Analyzing Expert Judgment: A Practical Guide,* Academic Press, London, 1991.

Skinner, D. C., *Introduction to Decision Analysis — Beginning Coursebook,* Probabilistic, 1996.

Stanley, L. T., *Practical Statistics for Petroleum Engineers,* Petroleum Publishing, Tulsa, OK, 1973.

CHAPTER 11

Nijhuis, H. J., and Baak, A. B., A Calibrated Prospect Appraisal System, *Proceedings of Indonesian Petroleum Association* (IPA 90-222), Nineteenth Annual Convention, 69-83, October, 1990.

CHAPTER 12

Campbell, J. M. Jr., Campbell, J. M. Sr., and Campbell, R. A., *Analysis and Management of Petroleum Investments: Risk, Taxes, and Time,* CPS, Norman, OK, 1987.

Davis, J. C., *Statistics and Data Analysis in Geology,* Volume 2, John Wiley & Sons, New York, 1986.

Van Horne, J. C., *Financial Management and Policy,* Prentice Hall, Englewood Cliffs, NJ, 1974.

CHAPTER 13

Ebdon, D., *Statistics in Geography — A Practical Approach*, Basil Blackwell, Oxford, 1981.
Skinner, D. C., *Introduction to Decision Analysis — Beginning Coursebook*, Probabilistic, 1996.

CHAPTER 14

Finkler, S. A., *Finance and Accounting for Nonfinancial Managers*, Prentice Hall, Englewood Cliffs, NJ, 1992.
Pariseau, R., and Oswalt, I., Using Data Types and Scales for Analysis and Decision Making, *Acquisition Review Quarterly*, 145-159, Spring, 1994.
Rodricks, J. V., *Calculated Risks — Understanding the Toxicology and Human Health Risks of Chemicals in our Environment*, Cambridge University Press, Cambridge, 1994.

Index

A

Absolute minimum and maximum values, 103, 104
Adding variables, 60-65, 104
Archiving, 200
Area under cumulative frequency curve, 161-166
Artificial intelligence, 89
Asking the right question, 57, 65
Axions, 90

B

Bar charts, 53-55, 108-110, 117
Bayes, Thomas, 79
Bayesian analysis, 79-82
Beta distribution, 126
Bimodal distributions, 53, 108, 112, 120-121
Bin sizes, 110
Brokerage house, 9, 27

C

Capital project risk model, 10
Cash flows, 19, 21
  common measure, 26
  comprehensive model, 207, 208, 220
  contributing-factor diagram, 64, 65
  time series analysis, 141, 142, 143, 146
Categories, 9, 10, 14, 48, 49, 56, 57
  discrete distributions, 119
  fate/transport model, 187
  specific distributions, 126
  qualitative model, 177
Chance of failure, 129-140
  Bayesian analysis, 81
  comprehensive risk model, 222-225, 227
  cumulative frequency plot, 101
  dependence, 141
  education, 69, 70
  risk process, 12
  risk-weighted values, 160, 168
Chi-squared distribution, 111, 126, 127
Circular dependence, 141, 152-153
Clipped distributions, 131, 133, 138
Combining Monte Carlo output curves, 104
Communication, 1, 3-4, 26, 35-40, 60, 75, 198

Comparisons, 91, 94-95
Confidence in most-likely value, 123, 124, 125, 179, 180, 211
Consensus model, 22-25
  education, 69
  locked models, 31
  risk process, 12, 19
  risk vision, 27
Consistency, 21-34
  checklist, 8
  double dipping, 60
  education, 69, 74
  legal model, 198, 199
  risk process, 12
Construction costs, 16
  comprehensive risk model, 207, 211-229
  consistency, 21, 22, 25, 32
  contributing-factor diagram, 55-58, 61-64
  education, 68
  time-series analysis, 181
  units, 48
Contributing-factor diagram, 51-66
  comprehensive risk model, 209, 211
  consensus model, 23, 24
  distributions, 110
  education, 69, 70
  legal model, 200
  risk process, 19
Correlation, 39, 69, 92, 149-153, 172, 173
Cumulative frequency curves, 95-100, 100-104

D

Decision trees, 82-87, 175, 192-198
Decisions, 105
  chance of failure, 136
  communication, 35, 36, 38, 40
  consensus model, 42
  decision trees, 82
  education, 68
  legal model, 198
  model types, 9
  risk process, 12
  time series analysis, 141
  updating estimates, 8
Define variables, 56-57, 65

Dendrites, 89-90
Dependence, 141-158
  decision trees, 198
  education, 69, 70, 71
  Monte Carlo analysis, 92
  risk process, 12
  sensitivity analysis, 171
  variable relationships, 5, 6, 7
Dependent variables, 147-149, 151-153
Deterministic analysis, 1, 3, 4, 7
  decision trees, 83, 192, 193, 197
  fate/transport model, 188
  Monte Carlo analysis, 91
  sensitivity analysis, 171
  time series analysis, 142
Discrete distributions, 105, 112, 119-120
Discriminant function analysis, 77-79
Distributions, 106-127
  beta, 126
  bimodal, 53, 108, 112, 120-121
  chi-squared, 111, 126, 127
  discrete, 105, 112, 119-120
  error function, 126
  exponential, 126
  flat, 105, 111, 112, 116-117, 119
  gamma, 110, 111, 126
  skewed, 53, 105, 108, 111-112, 114-116,
      120-121
  specific, 105, 126-127
  spike, 105, 111, 112, 115-116, 120, 121
  symmetrical, 105, 111, 112-114, 116, 117
  triangular, 105, 112, 125-126
  weibull, 126
Double dipping, 12, 51, 58-60
Drilling:
  oil well, 7, 29, 34, 39, 117, 123
  pollution, 6
  remediation, 79, 80
  water well, 133
Duration of exposure, 10, 188

E

Economic risk weighted resource value, 165-166
Education, 11, 13, 67-76
Effect of dependence on Monte Carlo output, 153
Eigenvalue, 88-89
Eigenvector, 88-89
Environmental compliance, 178-180
Environmental, health & safety models, 9, 187-192
Error function distribution, 126
Examples of risk assessments, 175-229
Expanding single values into distributions, 113,
      142, 143, 184, 217
Expert systems, 89

Exponential distribution, 126
Exposure model, 187-192

F

Facilitation, 41, 44-47
  comprehensive risk model, 205-209
  consistency, 23, 24, 27
  contributing-factor diagram, 56-59, 61, 64
  risk process, 11, 12, 14, 19
Factor analysis, 77, 87-89
Failure, 129-140
  Bayesian analysis, 81
  branching, 5
  categories, 49
  communication, 39
  comprehensive risk model, 222, 225, 227
  education, 69, 70
  Monte Carlo analysis, 101
  risk process, 12, 13
  risk-weighted values, 160, 168
  time series analysis, 141
Fate-transport model, 9, 10, 23-24, 77, 187-192
Fisher, R. A., 78
Flat distributions, 105, 111, 112, 116-117, 119
Frequency of exposure, 188, 190
Frequency plots, 53-55, 69, 95-100, 108-110

G

Gamma distribution, 110, 111, 126
Gausian distributions, 110, 112, 113, 123
Gosset, W. S., 91
Group dynamics, 44-45, 47

H

Health-based risk model, 9, 23-24, 187-192
Hierarchy diagrams, 46
High-level model, 16-17
Historical information, 121-123
  Bayesian analysis, 79-82
  chance of failure, 130
  consistency, 33
  dependence, 151, 152
  distributions, 111, 117
History, Monte Carlo analysis, 91-92

I

Identify variables, 56-58, 60-65
Independent variables, 147-149, 151-153
Influence diagrams, 51-53, 60-61, 64
Ingestion rate, 188
Intake, calculation of, 188

Iterations, 94-95
  chance of failure, 134, 136, 138-139
  consensus model, 47
  cumulative frequency plots, 102-103
  dependence, 147
  distributions, 122-123
  risk-weighted values, 161-162
  time series analysis, 182, 187
Internal rate of return, 10, 25, 28, 42, 61, 71,
    142

K

Kolmogorov-Smirnov test, 95
Kurtosis, 111, 123-125

L

Latin hypercube, 122
Learning, examples:
  Neural nets, 89-90
  Reinforcement, 90
  Supervised, 90
  Unsupervised, 90
Legal applications, 9
  communication, 4, 36, 198-204
  distributions, 110
  education, 72
  legal risk model, 198-204
  risk vision, 27, 28
  time series analysis, 144
Linear regression, 151
Loadings, 89
"Losers," 12, 15, 38

M

Magnitude of coefficients, 170-172
Make-it-bigger philosophy, 16-17
Management, role in:
  comprehensive risk model, 207, 208
  consistency, 22, 23, 32
  contributing-factor diagram, 53
  decision trees, 85
  education, 67-69, 71
  Monte Carlo analysis, 104
  risk process, 15
  time series analysis, 144
Maximum distribution value, 111-112
Mean values in:
  consistency, 30
  decision trees, 85
  distributions, 111, 112, 122, 123
  Monte Carlo analysis, 94-95, 104
  risk-weighted values, 164

  sensitivity analysis, 171
  time series analysis, 143
Minimum distribution value, 111-112
Mitigation, 37-38, 45, 174, 220-221, 223
Model types, 9-10, 175-229
Monte Carlo analysis, 91-104
  chance of failure, 133, 134, 136, 139
  comprehensive risk model, 212
  decision trees, 87, 193, 195, 196, 198
  dependence, 147-148, 152, 153
  distributions, 116, 119, 120, 122-123, 125
  education, 69, 70
  neural nets, 90
  risk assessment advantages, 2, 6
  risk technologies, 77
  risk-weighted values, 161-162, 170
  sensitivity analysis, 171
  time series analysis, 143, 144-147, 183
  qualitative model, 176, 177
Most-likely distribution value, 111-112
Multiple dependencies, 152-153

N

N-dimensional space, 77, 78
Negative correlation, 149, 173
Net present value, applied in:
  consensus model, 42, 46, 47
  consistency, 28, 31
  contributing-factor diagram, 61
  education, 70, 71
  risk advantages, 2, 10
  risk process, 15
  time series analysis, 141, 187
Net risk weighted value, 164-165, 168, 220, 227
Neural nets, 77, 89-90
Neuron, 89-90
New information, 79, 80
New-products risk model, 10

O

Operating expenses applied in:
  consensus model, 48
  contributing-factor diagram, 57, 58, 62
  decision trees, 82-87
  Monte Carlo analysis, 103
  time series analysis, 181-184
Organizational changes, 14-16, 71-72
Overarching categories, 48, 49

P

Pattern recognition, 90
Peakedness, 111-112, 114, 123-125

comprehensive model, 211-212, 216-217, 222-223
mini time-series model, 182-186
semi-quantitative model, 179-180
Pipeline risk model, 10, 48, 55-56
  capacity, 121
  construction, 57
Poisson distribution, 110
Political considerations, 6, 10, 11, 13, 14
  categories, 49
  common measure, 25
  consensus, 42, 44
  consistency, 30
  contributing-factor diagram, 55, 65
  decision trees, 85
  defining question, 41
  education, 73
  recording, 46
  risk-weighted values, 164
  sensitivity analysis, 168, 169, 173
  time series analysis, 141
  units, 48
  zoning, 222
Political risk model, 10
Porosity, 8, 92
Portfolio management applied in:
  communication, 38
  consensus model, 44
  consistency, 24, 25, 26-28, 31
  distributions, 126
  Monte Carlo analysis, 95
  risk advantages, 1, 2, 8, 9, 10
  risk process, 15, 16
Posterior probability, 78-79
Prediction, 90
Preparation of organization, 14-16
Pre-project estimates, 33, 58
Principal components, 88
Prior probabilities, 79, 80
Probabilistic analysis, 36, 167
Probability segments, 98-100
Problem types, 9-10, 175-229
Process, risk assessment, 11-14

Q

Qualitative models, 30, 175-177
Question, the, 41-42, 57-58, 65

R

Random comparisons, 94-95
Rank-ordering answers, 29, 175
Reading data from files, 121-123

Regression analysis, 151-152
Reinforcement learning, 90
Remediation, 6, 23-24, 79-80
Risk assessment process, 11-14
Risk-weighted values, 159-168
  common measure, 25
  comprehensive model, 220, 227
  cumulative frequency, 101
  education, 67, 69-70
  factor analysis, 89
  legal model, 202
R-squared value, 151

S

S-shaped curves, 100-101
Security, 150-151, 176-177
Semi-quantitative models, 177-180
Sensitivity analysis, 168-173
  education, 69-70
  fate/transport, 192
  legal, 199, 202-206
Signal processing, 90
Skewed distributions, 114-115
  bimodal distributions, 120, 121
  contributing-factor diagram, 53
  flat distributions, 116
  how to approach distributions, 111, 112
  what a distribution is, 108
Sludge pits, 81-82
Sorting, 19, 46-47
  chance of failure, 134
  education, 70
Spearman rank correlation, 172
Statistics-free risk assessment, 87, 111
Stochastic analysis, 4, 9, 11-13, 16
  communication, 36
  decision trees, 193
education, 67
  fate/transport, 188
  reading from a file, 122
  time series analysis, 141
Students T test, 91, 112
Subdividing variables, 16-17, 61-65
Supervised learning, 90
Symmetrical distributions, 105, 111, 112-114, 116, 117
Synapses, 89-90
System modeling, 90

T

Tailed distribution, 114, 117, 125-126
Target classification, 90

Time-series analysis, 141-147
  comprehensive model, 216-217, 222
  construction, 64
  decision trees, 83, 85, 87
  education, 70, 71
  model, 180-187
  pipeline model, 10
  updating estimates, 9
Timing, 24
Tornado diagrams, 170-171, 172
Total chance of success, 80-81, 129, 134, 227
Transfer function, 90
Triangular distributions, 105, 112, 125-126
Truncated distributions, 112, 117-119, 121, 126, 184

U

Uncertainty, 1, 5
  contributing-factor-diagram building, 51, 61
  decision tree, 193
  decisions, 105
  distributions, 106
  education, 68, 70
  fate/transport, 188
  legal, 199

Uncorrelated variables, 88, 149
Uniform distributions, 116
Units, 28, 47-48
  chance of failure, 134
  comprehensive model, 211
  contributing-factor diagram, 53, 55, 64-65
  economic risk-weighted resource value, 165
Unsupervised learning, 90
Updating estimates, 8-9, 80

V

Variance, 88-89, 111, 149, 173

W

Water treatment, 121
Weibull distribution, 126
Weights, 10, 120, 176-177, 180
"Winners," 12, 15, 38
Wording of questions, 56

Z

Z-shaped curves, 100
Zoning, 222, 225, 227